Society, Culture and the Auditory Imagination
in Modern France

# Society, Culture and the Auditory Imagination in Modern France

## The Humanity of Hearing

Ingrid Sykes

*Research Fellow, La Trobe University, Australia*

First published 2015 by
PALGRAVE MACMILLAN

Palgrave Macmillan in the UK is an imprint of Macmillan Publishers Limited, registered in England, company number 785998, of Houndmills, Basingstoke, Hampshire RG21 6XS.

Palgrave Macmillan in the US is a division of St Martin's Press LLC, 175 Fifth Avenue, New York, NY 10010.

Palgrave Macmillan is the global academic imprint of the above companies and has companies and representatives throughout the world.

Palgrave® and Macmillan® are registered trademarks in the United States, the United Kingdom, Europe and other countries.

ISBN 978-1-349-49813-0      ISBN 978-1-137-45535-2 (eBook)
DOI 10.1057/9781137455352

This book is printed on paper suitable for recycling and made from fully managed and sustained forest sources. Logging, pulping and manufacturing processes are expected to conform to the environmental regulations of the country of origin.

A catalogue record for this book is available from the British Library.

Library of Congress Cataloging-in-Publication Data
Sykes, Ingrid.
   Society, culture and the auditory imagination in modern France : the humanity of hearing / Ingrid Sykes.
      pages   cm
   Summary: "This book examines the striking way in which medical and scientific work on hearing in eighteenth and nineteenth-century France helped to shape modern French society and culture. Contemporary scientists and anatomists had to come to terms with a new kind of transformative physiology within the material site of the human ear, one that had the potential to construct space and place in the most powerful way imaginable. Auditory medical specialists found themselves at the center of pivotal philosophical, political and social debates on how the individual citizen might use their ears to reach out to those around them constructing broader, protective models of social reform. Sykes makes the case that of all the senses, hearing offered the greatest resources for remodelling the idea of the universal human condition within the modern French historical setting" — Provided by publisher.

   1. France—Civilization.   2. National characteristics, French.   3. France—Social life and customs.   4. Auditory perception.   5. Listening—Social aspects—History.   6. Listening—Political aspects—History.   7. Listening—Psychological aspects—History.   8. Physicians—France—Attitudes—History.   9. Judges—France—Attitudes—History.   10. Empathy.   I. Title.
   DC33.8.S95 2014
   944—dc23                                                    2014026291

# Contents

# Figures

# Preface

Tracing a history of hearing is not straightforward. It requires that we consider not simply the sounds or sonic fields of the past but the broader imaginative models that hearing encompassed at that particular time and place. Throughout French modern history, hearing was a fundamental part of the way in which humans thought about themselves and those around them. Embedded within its meaning was not simply the activation of a specific physiological human "sense" but a specific stance or posture that caused the redefinition and reshaping of social, political and cultural boundaries and their spaces. Hearing, from this perspective, went well beyond auditory languages and landscapes of information to encompass those discourses and spheres emphasizing critical forms of social and political care and protection: the courtroom, the hospital and the disabled schoolroom. Hearing might involve hearing actual sounds in these places, but it was not ultimately dependent on them. Rather, hearing in these sites primarily occurred through the positive activation of the human imagination.

Implicit in the imaginative hearing model was a paradox – a crisis, perhaps – that positioned the individual at the crossroads of their own individual sense of self and the collective social sphere. How can we exist as ourselves whilst coexisting with others at the same time? Yet hearing offered powerful resources in resolving this paradox. To hear was to confront the fragile edges of the self and in turn free one from oneself to forge a positive relationship with others. It is easy to overlook such a radical self/subject framework. Indeed, as we rely increasingly on subjective experience, abstract pools of information and the visual field, hearing, aside from acting as a basic mode of cognition, appears almost obsolete. Yet in recent public and scholarly discourse, particularly in the medical and public health field, consideration for hearing and listening in this more deep, holistic sense is gaining ground. Doctors are now asked, for example, to listen more carefully to patients. Alternative medicine is gaining mainstream recognition. Modern medical practice involves not only thinking about a patient's problem but also imagining how they might feel and acting appropriately. There is an increasing sense that, without evidence of a sophisticated hearing stance, the modern medical industry cannot purport to care for people or understand their needs.

This book brings the historical narrative of this hearing stance into greater focus. It reveals that hearing in France played a formative role in shaping modern society as we know it today. Significantly, the book focuses a great deal on the role of medicine within this process. Just like medical practitioners today, those in French modern history were concerned with how they were hearing their patients. The collision of medical otology and French philosophy during the Enlightenment initially forced hearing as an empathetic social practice into the public domain. Prior to the revolution, the French élite again consolidated hearing's presence at the forefront of social change by envisaging radical spaces of hearing, such as the medical clinic, the courtroom and the modern hospital. They reinforced the broader political message that the people must be heard. During the post-revolutionary period, the body, as a whole, took on a hearing stance. The auditory body was a dynamic hearing posture in which the individual self could attach itself to a collective form of state care. Technologies became increasingly important in this listening act, fixing the individual listener to his/her own intimate communicative framework.

To study such a history of hearing is not to undermine the significance of music, art or science, disciplines which intersect with hearing in important ways. Rather, it is to reinterpret and highlight these disciplines and their languages as key tools in contributing to a more powerful social metalanguage that had the capacity to create change entirely on its own. Significantly, medical and judicial institutions, where techniques and practices of hearing were most evident, became central in channelling and shaping this agenda. In the clinic, pulse theories reinvented musical idioms in order to connect patient to doctor. Industrial materials fashioned linguistic devices so that the blind could hear messages of belonging in their own institutional spaces. The modern hospital and the small claims courtroom emerged as one of the most sensitive listening sites. This was where the cry of "humanity" was loudest and most intense.

This book does not chart a straightforward chronological history. It suggests through a variety of case studies the ways in which hearing might have shaped modern French society from the Enlightenment to the modern industrial age at its deepest level. Hearing becomes an open and closed mode of topographical analysis, an anthropological tool for uncovering relational patterns which became set in the landscape to form our modern, social settings. Such a process of activity, like hearing itself, was inherently dynamic. Shifting voices created shifting spaces which might succeed or fail. The French Revolution can be interpreted as a heightened point of intensity in society's hearing flow. It represents the fragile point in the hearing process where perceptual gambling had to begin. The crisis of the "false noise" consumed the entire social and political setting for a considerable period of time until the hearing process could resume and the social setting stabilize. Hearing as a positive practice in constructing human relationships has continued to feature as a key defining cultural, political and social force within the modern world.

# Acknowledgements

I extend my special thanks to Holly Tyler at Palgrave Macmillan for bravely taking on this project for publication. This book could not have been completed without the support of a major research fellowship from the Wellcome Trust, UK. The Centre for the History of Medicine, University of Warwick, UK, generously hosted this fellowship and supported my research ideas from the very beginnings of the proposal process through to publication and completion.

Thank you also to Daniel Pressnitzer of the Equipe Audition, Laboratoire Systèmes Perceptifs, Ecole Normale Supérieure, Paris. His highly imaginative work on auditory scene analysis, which he presented as a keynote at my 2009 Warwick international symposium, "Signalling Sound", helped facilitate key conceptual shifts in my thinking that led to the formation of this work. Another important individual whom I must thank is Penelope Gouk, who shared her expertise with me during this lengthy research period in such a warm, generous and spirited fashion. Her enormous breadth of knowledge on seventeenth- and eighteenth-century European sound culture as well as her skill in analysing medical treatises from this period has influenced my writing and thinking on this topic in a profound and lasting way. To those who read my drafts and provided valuable input, Ian Coller, Daniel Grimley Ida Nursoo, Susan Sykes and the highly intuitive students from my 2013 La Trobe University Honours Group, "History and the Human Condition" – thank you all for your helpful insights and comments on my work.

This work was supported by the Wellcome Trust, UK (75002).

# Introduction

During the late seventeenth century, French philosophers were beginning to think seriously about hearing. If an individual could hear with a moral purpose in mind, society as a whole could not fail to remain good. At this stage in history, hearing was about connecting to another reality, one of higher purpose. Perceiving another person's words was not enough. Hearing was about focusing oneself on moral transformation, rising above the bad emotions associated with the dangerous voices of an increasingly secular world. Only then could society be improved. In Francois Fénelon's 1699 epic novel, *Les aventures de Télémaque*, for example, the hero encounters a number of different sounds on his voyage to become king.[1] At the beginning of the story, Fénelon deposits Telemachus, accompanied by Minerva, the Goddess of Wisdom, in the guise of his tutor Mentor, on the isle of Calypso. Here he confronts his first listening challenge. Calypso lived in a wonderful grotto, scooped out of the rock in arcades abounding with pebbles and shell work, extending into hills and clouds like a great garden. In need of refuge and kindness, the vulnerable Telemachus hears the murmuring fountain, the song of birds and the sound of the brook. Calypso's nymphs begin to sing, and Calypso starts to speak. Telemachus succumbs to his passions. Another voice, however, warns him of possible deceit: "Take care how you listen to the soft and flattering speeches of Calypso, which will glide like a serpent under flowers; dread that concealed poison; be diffident of yourself, and never take any resolution without first waiting for my advice."[2] Mentor's voice contains the power to direct Telemachus away from the isle. On board the ship commanded by Adoam, Telemachus hears a new set of sounds, Mentor's artful lyre playing and his perfectly nuanced singing voice.[3] This is accompanied by Adoam's description of Bétique, an ideal rural community governed by the experience of

1

the ancient sages and the uncommon wisdom of some young men. In the remainder of the story, Mentor saves Idomeneus' kingdom from disaster and prepares Telemachus for the role of king. He warns both Idomeneus and Telemachus against those corrupt qualities, pride, ostentation, voluptuousness and inhumanity, that might lead a kingdom to ruin. At the end of the story, Mentor reveals her true identity as the goddess Minerva and gives her final words to Telemachus: "When you ascend the throne, let the great object of your ambition be to renew the golden age. Let your ears be open to everyone, but let your confidence be confined to a few... Love your people, and neglect nothing that may tend to conciliate their affection."[4]

Throughout this story, Telemachus listens in order to engage his auditory imagination. His listening attends to sounds that led him into peril, as represented by the temptations of the isle of Calypso. Predominantly, however, he attains that ultimately moral state of being which can direct his own actions towards the social good. This is depicted by the sounds of the idealized world of Bétique and the wise voice of the goddess Minerva. By the end of the story, Telemachus demonstrates that he has indeed affected a life-long listening relationship with others by showing personal responsibility as leader. Most of all, Telemachus reveals his ability to "care for" and "love his people". The ability to bond with people through an imaginative hearing process is a central part of Fénelon's novel. Such a theme drives the entire narrative of Telemachus' transformation from vulnerable traveller to ideal ruler. The imaginative way in which Telemachus attains such a moral relationship with those around him was, of course, much like the Biblical stories of Creation. It was the constructive power of God's imagination that fashioned the divine universe. This power was replicated in man. It was only when Adam in the Garden of Eden had the capacity to use his inner self to imagine that he could begin to take full responsibility for his place in the world.

In *Reason and Resonance* Veit Erlmann has recently drawn attention to the eighteenth-century ear as a complex space of subject/object discourse.[5] The question of "Who are we?" was directly informed by the listening subject. He connects anatomical interest in the ear amongst seventeenth-century Italian anatomists directly to the Foucauldian notion of "duplicated representation", a "bond established, inside knowledge, between the idea of one thing and the idea of another".[6] But Erlmann also suggests that philosophers used the space to stage a fragile balancing act between the reflexivity of sound perception and the autonomy of mechanical processes. Significantly, he reveals that Descartes explored the ear not simply as a passive sensory reflex but as a more constructive and creative human tool used to explore "self" formation. As Erlmann writes:

> More than the effect of the philosopher's alleged youthful cognitive ambivalence displayed in the Compendium or the aesthetic ratio-nalism of his mature period, Descartes' theory of hearing was one of the central sites of his lifelong quest for the great entente of rea-son and emotion. Grappling with the impossible proximity of reason and resonance, he simply gave the ear a chance to cast its lot with either.[7]

Erlmann's study supports other scholarship that draws attention to the increasing emphasis on psychological self-sufficiency amongst cit-izens surrounding eighteenth-century cultures of knowledge. In *The Post-Revolutionary Self* Jan Goldstein draws attention to a more fragile form of self-operating within and around the culture of inwardness that was more unstable. She writes: "One such subnarrative... is the concern about a specific quality of psychological inwardness: whether the inner space that is inhabited enjoys a structural unity that can ground a reliably coherent self or is so fragmented and diffuse as to preclude that outcome."[8] Goldstein demonstrates that institutional set-tings provided spaces where the fully formed psychological self could be recognized and practised. Both Erlmann and Goldstein understand the eighteenth-century self as a kind of fractured entity that might be made whole through an externally directed, constructed, knowledge based, sensory experience. Erlmann's enlightened ear forges its own kind of fulfilment through an unstable balance of reason and emotion embed-ded in the Cartesian or, as he also demonstrates, the animist listening act.[9] Goldstein's unstable imaginative self (attending to "the perils of the imagination") requires stability through an encounter with a particular kind of institutional training based on the principles of sentiment and sensationalism. The eighteenth-century historian is constantly aware of how the eighteenth-century self might fight against "it-self" through some physical or psychological means in order to be made whole. At the heart of this fracture, I contend, is a further disjunction between the interiorized nature of the eighteenth-century self and its ability to reach out beyond the individual interior. How did the eighteenth-century individual even begin to become socially aware when it was so solely focused on constructing and re-constructing its own inner self?

Sophia Rosenfeld's book *A Revolution in Language* confronts this ques-tion by acknowledging the powerful role of language, gesture and communication in the construction of social and political ideals during the Enlightenment and the French Revolution.[10] However, it is not sensory perception per se which is the focus of Rosenfeld's thesis. Rather, it is the way in which sign, as articulated through gesture, is used to resolve larger questions surrounding words and language in the eighteenth century. She argues that the same utopian ideals surrounding language that existed prior to the revolution ultimately shaped revolutionary preoccupations and anxieties with *la parole* ("the word") as well as inflecting post-revolutionary debates over race and law. By shifting the focus away from sensory activity as an expression of inner fragmentation, Rosenfeld demonstrates its powerful impact on the cultural landscape as well as on major political and social transformations such as the French Revolution. Rosenfeld has also recently emphasized that "being heard" was essential during the revolution. She explains that "true citizenship entails becoming a full participant in the intersubjective game that is politics. Furthermore, this activity involves not only getting to speak (i.e. having 'a say' or 'a voice'), but also being actively and attentively heard."[11] Hearing itself has also gained significant attention amongst other eighteenth-century Enlightenment historians. The idea of the "Enlightenment as Conversation" has been visited and revisited by Jürgen Habermas, Lawrence Klein, Dena Goodman and others at length. Such historians locate the medium of the Enlightenment "not in light and vision but in sound and speech. In this Enlightenment, engaged conversers rather than detached observers are in the foreground."[12] In his anthology, *On the Pragmatics of Communication*, Habermas outlined the parameters for a socializing "public" sphere that might be understood solely through speaker/hearer agreement.[13] Goodman presents the Enlightenment through a similar kind of interchange. However, she deliberately focuses on the interior world of the French seventeenth- and eighteenth-century salon as the space for its cultivation: "If men of letters learned to defend their honour in aristocratic salons in the seventeenth century, they also learned that the formalized, rule-bound discourse was the best way to overcome the problems that social mixing entailed."[14] Goodman suggests that women *salonnières* were essential in establishing and maintaining a formal mode of discourse helping to formulate the "public" sphere. They stood apart from the conversation but intervened when an argument amongst male speakers veered out of control. The way in which Goodman examines conversation leads her to question the usefulness of talking about "private" and "public" spheres during the eighteenth century. In her important article on the topic, she writes:

The eighteenth century was the historical moment in which public and private spheres were in the process of articulation, such
that no stable distinction can or could be made between them – a
moment in which individuals needed to negotiate their actions, discursive and otherwise, across constantly shifting boundaries between
ambiguously defined realms of experience.[15]

Such eighteenth-century historians acknowledge the role of hearing
in constructing eighteenth- and nineteenth-century culture, yet what
is it about hearing that is so powerful and how exactly did it figure
in the development of modern France? Was its role merely to create
codes of agreement or is there something about hearing in particular that might be present in the Telemachus text? There have been a
number of more deliberate case studies on the history of sound perception in French culture that are more specific about the nature of
audition within society. They point to the use of sound within French
historical culture as a vehicle for social inclusivity. These case studies demonstrate that sound was not simply a means to an end in the
creation of "civilising languages", commerce or technological prowess
during the Enlightenment and the nineteenth century. Rather, it was
a fundamental and guiding source of social cohesion. James Johnson's
examination of listening at the Paris Opera during the late eighteenth-
century demonstrates the power of sound to create public ritual and
political sentiment.[16] He explains that the Parisian elite went to the
opera primarily to feel and experience together specific kinds of sentiments through the sound of the music and the words of the text.
Political circumstances governed the ways in which the individual
responded to the music that was presented to them, yet they all practised hearing together in a public setting. Johnson demonstrates, like
Goodman, that the public and private interacted at the Opera through
the listening experience in a very complex and fluid way. The space
of the opera was critical in establishing the individual listener at the
centre of a very collective public zone of experience. But the individual was also alone, to a degree, in a private world of feeling and
perceiving. Alain Corbin's study, *Village Bells*, shows that it was the
capacity of the village bell to make people "foregather" through sound
that made it so highly sought after by various political factions within
society.[17] Corbin charts the history of a society that was driven not by
the circulation of print or written texts, press releases or caricatures, but
the single, penetrating sound of the village bell ringing out across the
landscape.

Hearing cultures in historical modern France were not, however, limited to such obviously acoustic examples of collective, social expression. An emphasis on hearing and place was explicitly articulated by late seventeenth-century philosophers who advocated a new ethical awareness of hearing generally as a resource for connecting with others within a shared public zone of existence. Traditional models of hearing which had previously occurred in the confessional box, the bell tower or the church ritual, was now advocated in a number of different new and existing secular spaces, where it retained the same transformative character. In addition, technologies appeared which were deliberately constructed to encourage the encoding of a particular space through sound. Such cultural artefacts were much more than systems of sensory signs designed for observation or knowledge. They were expressions of a culture that remained reliant on hearing as an ethical, transformative experience knitting the private individual with the public space.

French surgeons were pivotal to the development of this culture of the auditory imagination. The anatomist Joseph Duverney (1648–1730) used written text and illustration so as to articulate the transformative implications of hearing within the *site* of the human ear.[18] The development of spaces, bodies and technologies during the eighteenth and nineteenth centuries were simply extensions of this kind of scientific thinking. Hearing supplied the extended spatial and temporal field used to imagine and conceive them. So immersed was the individual in this auditory world that he/she was able to access others. Erlmann has touched brilliantly on this quality of sound in eighteenth-century culture, in his discussion of Claude Perrault's response to Robert Cambert's opera, *Les peines et les plaisirs de l'amour* (1672). "Listening to the *bruit agréable* of Cambert's pastorale here comes to be understood in the larger sense of the ear becoming absorbed by its own potential for the infinite expansion of the audible – as the possibility of gaining untold access to an unheard, yet eminently representable world."[19] But Erlmann, amongst others, has not endeavoured beyond the pure bourgeois musical experience to explore the broader social consequences of this form of auditory experience.

I argue that during the eighteenth and nineteenth century in France, hearing was an implicit feature of many spaces, sites, bodies and mechanical technologies within society and culture. Human hearing became a critical vehicle for the spatial transformation of the social world and for the coding of that social world. Social change during this period relied on a transformative model of sensory experience that connected the individual self to a broader realm. Prior to the eighteenth century, hearing connected man to the cosmos via a complex series of mathematical relationships. French surgeons then simply transplanted the realm of hearing from its abstract cosmic and mystical domain (experienced in the religious areas of church and cathedral) to the site of the human ear and certain social spaces of society where audition was cultivated. What changed was not the model of hearing but the philosophical vocabulary surrounding it. French society remained dependent on an auditory model from the religious past because it offered dynamic possibilities of transformation that broadened the self to collective realms of space and place. Such models became critical to the invention of new sites of power that were marked by a more ambiguous private/public interface of construction. Rather than examining bourgeois musical works, I focus on spaces and places which relied primarily on sound to make their focus on "public" inclusivity work within a private institutional or authorial domain. Many of these cultural features fostered the auditory imagination to resolve social hierarchies at the same time as consolidating an existing authority. The nuanced model of the individual who engages with his auditory imagination as represented in the figure of Telemachus, relied on sound's potential to enact exactly what vision cannot: to cultivate audition as a means of situating the self within a broader collective site of social experience.

## The significance of hearing in modern France

There is a growing band of work demonstrating the unique capacity of sound as opposed to the other senses to influence culture and to shape society. Jonathan Sterne, Michael Bull and Les Back, Mark Smith and Veit Erlmann have put together excellent collected volumes of essays which draw together scholars from a number of different disciplinary fields with the aim of consolidating the field of sound studies.[20] Sound studies, as defined by Sterne, is very different from academic fields such as linguistics or music analysis which study "pure sound". It is an interdisciplinary avenue of scholarship that takes "sound as its analytical point of departure or arrival. By analysing both sonic practices and the discourses and institutions that describe them, it redescribes what sound does in the human world and what humans do in the sonic world."[21] These collected editions touch on such varied subjects as the history of European soundscapes, listening and social media, and noise abatement campaigns in modern America. Sound studies attempts to demonstrate the centrality of sound in global society by comparing and contrasting the different ways in which sound constructs knowledge production, power and culture. Such a field has emerged as a powerful challenge to the "Othering" of sound against vision both in society as well as in academic discourse.

Concern over the primacy of vision was also a central preoccupation of the work of Walter Ong, Jesuit priest and professor of English literature, religious studies and culture. With the publication of *The Presence of the Word* in 1967 Ong made one of the most important contributions to formulating a sound theory for culture and society touching on its meaning in both the East and the West.[22] For Ong, oral cultures have specific characteristics that are linked to the nature of hearing. Since sound is produced in a present moment, it relates power, force and dynamism where vision cannot. It establishes a "here-and-now" kind of presence – the "voice". Oral cultures create inclusiveness using hearing's special relationship with interiority: "Sound has to do with interiors as such, which means with interiors manifesting themselves, not as withdrawn into themselves for true interiority is communicative."[23] The sound signal is often an indicator of action or movement and unites and destabilizes groups of living beings through these features. It creates a presence for living beings to inhabit. Sound is linked to thinking, which is an interiorized activity linked to voice. Like gesture, sound manifests from the interior human body but cannot be frozen in time. He explains:

> Vision, as stated earlier, of itself manifests only surface, superficies, outsides; to see the inside of anything, one has to make somehow an outside, cutting the object open or entering it (as a room) to see its interior as a surface. Exploring an interior by touch violates the interior, invades it, as we have likewise seen.[24]

Ong makes a strong distinction between the enforced spatialization of sound which is the word on the page and the spatialized sensibility that sound can create in and of itself. He writes: "It is not spread out as a field in front of us as a field of vision but diffused around us."[25]

Ong's work is underpinned by a broader argument that society became fundamentally different with the introduction of written culture. Once the word became spatialized on the page with the introduction of alphabet, its power as a "happening" was eroded. This crisis reached its height during the eighteenth century, Ong argues, the great age of printed texts. In her groundbreaking book, *Music, Science and Natural Magic in Seventeenth-Century England*, Penelope Gouk challenges Ong's argument that vision replaced sound in a simple "once-and-for-all shift" by demonstrating that English culture relied on non-verbal modes of discourse to construct scientific concepts.[26] One of the most important conclusions of her detailed study on music, natural magic and experimental philosophy is that models of musical resonance offered many important English scientists such as Newton and Hooke the resources for demonstrating through mechanical experimentation the laws of the invisible world of Creation. Gouk demonstrates that there was no process of "devocalization" as this culture of scientific experimentation took hold. On the contrary, society remained reliant on music's proximity to natural magic and the occult in order to cultivate the mathematical languages of modern physics. As she explains: "The experimental practice first made fashionable in Interregnum Oxford, and then successfully established by the Royal Society, *simply took over* assumptions and procedures which had previously been identified as natural magic. What changed was the form of rhetoric used to justify their goals."[27]

Gouk demonstrates the important contribution made to this Europe-wide intellectual culture of music, science and the occult by French acousticians, in particular, the work of the French Minim Friar Marin Mersenne. Mersenne, who collaborated at points in his career with Descartes, has made major discoveries in relation to modern acoustical law. Yet Gouk demonstrates that it was his commitment to the idea of "universal harmony" grounded in Catholic conceptions of the "Divine Wisdom of the universe" that facilitated his mathematical work. In Mersenne's world, laid out in his enormous treatise, *Harmonie Universelle* (1636), the rational language of science was merely the "trace" of a much deeper, spatial construction of the Universe that could not be expressed in words. Mersenne's insistence that it was a moral duty to acknowledge such divine laws of harmony paved the way for a powerful French philosophy of human hearing taken up by Catholic philosophers towards the end of the seventeenth century. Though they cultivated different strands of Catholic thought, Blaise Pascal (1623–1662), Nicolas Malebranche (1638–1715) and François Fénelon (1651–1715) all advocated a much greater attention to the act of hearing as a manifestation of divine wisdom. They explained in their writings that humans were failing to hear in a proper deep sense. The consequences were that individuals had no sense of their broader place in the world and thus their ability to behave with compassion towards others was hampered. Such philosophers advocated ethical or moral hearing outside the confines of the Church or convent paving the way for new social arenas within which this act might take place.

Part of the problem in tracing such a history of this cultural mode of hearing is our insistence on presenting and re-representing the French Enlightenment (and the "age of Modernity" in France at large) as a period which broke entirely from the religious past. Sound studies scholars such as Jonathan Sterne have persuasively confronted the primacy of vision in relation to this period in general (even describing it as a period of "Ensoniment"[28]) yet has remained sceptical to discussions of sound which give attention to "interiorizing" models which have evolved from Christian theology. This glosses over the French Catholic context from which modern French hearing cultures have evolved (whether they rely on an openly "secular" vocabulary or not) and the mystical (or "ill-posed" as a mathematical problem) way in which sound works in French culture. It was the adaptability of the mechanical spiritual model to map onto social, spatial needs of the new age that made it such an important resource for French scientists and anatomists. French modern auditory spaces during the eighteenth and nineteenth centuries were (and still are) necessarily radical in their modes, materials and technologies because the concept of "voice" or "heavenly harmony" within the French Catholic tradition is fundamentally imaginary (otherworldly), therefore also often materially synthetic. There is no distinction between the "voices" of nature and the mechanical "voices" of technology within this very distinctive cultural context and tradition. I touch on auditory sites which demonstrate an array of different sounding materials: the courtroom (verbal complaints), the medical clinic (cries, arteries), the blind school and body (music, word, technological), the individual body (technological). In addition, I emphasize the way in which French inventors developed and cultivated a culture of "communication objects", technologies which manufactured systems of "voices" in a similar way to the enormous mechanical objects within French Catholic spaces of the pre-Enlightenment past.[29]

The French cultural hearing experience was most powerfully articulated by Jean-Luc Nancy in 2002.[30] Here, Nancy describes a form of hearing or "listening" (which in French, *écouter* is different from *entendre*, to understand) which facilitates access to a broader realm through the idea of the sonorous present:

> For this reason, listening – the opening stretched towards the register of the sonorous, then to its musical amplification and composition – can and must appear to us not as a metaphor for access to the self, but as the reality of this access, a reality consequently indissociably

"mine" and "other", "singular" and "plural", as much as it is "material" and "spiritual" and "signifying" and "a-signifying."[31]

Nancy embeds listening within the idea of the "sonorous present", a concept, like Mersenne's idea of the divine harmonic universe which carves out a space by altering normal conceptions of time: "The sonorous present is the result of space-time: it spreads through space, or rather opens a space that is its own, the very spreading out of its resonance, its expansion and its reverberation."[32] Such an idea of hearing, Nancy explains, is not embedded in real spaces but in an otherworldly conception of space once hearing takes place: "So the sonorous place, space and place – and taking-place – *as* sonority, is not a place where the subject comes make himself heard (like the concert hall or the studio where the singer or instrumentalist enters); on the contrary, it is a place that becomes a subject insofar as it resounds there."[33] Finally, Nancy reiterates the proximity of the listening subject to "voice" within the resonant place which bears exactly the same characteristics as the Catholic divine. He writes: "[A] 'voice' we have to understand what sounds from a human throat without being language, which emerges from an animal gullet or from any kind of instrument, even from the wind in the branches: the rustling toward which we strain or lend an ear."[34]

France has maintained a rich tradition of exploiting hearing as an imaginative transformative tool throughout the more recent past. Throughout the latter half of the twentieth century, French writers, musicologists and composers promoted sound as a vehicle for collective social coding in and of itself. The most obvious scholarly example is Jacques Attali's work, *Noise* (1977), which charts music's relationship with economic realities: "More than colours and forms, it is sounds and their arrangements that fashion societies."[35] Attali traces the way in which music mirrors and forecasts major social changes, demonstrating its simulation with the ritualized practices of the pre-Enlightenment era through to the primitive capitalism of the late eighteenth and nineteenth centuries and onto the realities of twentieth-century mass production and social organization. Just as important were the compositional systems and musical techniques developed by Pierre Schaeffer and later those at the famous Parisian electro-acoustic institution Institut de Recherche et Coordination Acoustique/Musique (IRCAM).[36] A whole generation of musicians began to subject sound materials ("sound objects" as they called them) to complex technical manipulation so as to fashion an imaginative world of sonority for the individual to experience. Michel Chion argued that visual media such as film could only be made truly expressive through careful attention to sound. In the place of the musical score was something called the synchronized "sound track" that relied on a whole variety of audio techniques to transform the visual material into something frighteningly confronting. There is currently scientific research occurring within France in the field of auditory perception analysis which has also been shaped by this cultural tradition to some degree. Scientists working in the hearing perception field have been examining and charting in much more detail the patterns and codes through which the listening subject connects with the complex acoustical environment surrounding the human ear and its connecting neural field. French scientists such as Daniel Pressnitzer, for example, now construct highly creative models of analysis incorporating the complexities and intricate nuances of the auditory scene. Such scientists confront the way in which the human brain must make sense of the space-time "transduction" within the auditory resonant domain.[37]

The French cultural attitude towards sound and hearing/listening (I use these terms interchangeably) as a transcendental extension of the self has also influenced the work of important French philosophers including Michel Foucault.[38] There has been a great deal of emphasis from the Anglo-speaking world on the "gaze" in a literal visual sense within his historical work. Yet within other areas of his output, Foucault reveals a keen interest in the way in which hearing might work to construct the visual image.[39] In his introduction to Baudelaire's *The Temptation of Saint Anthony* he writes, for example:

> Dreams are no longer summoned with closed eyes, but in reading; and a true image is now a product of learning: it derives from words spoken in the past, exact recensions, the amassing of minute facts, monuments reduced to infinitesimal fragments and the reproductions of reproductions. In the modern experience, these elements contain the power of the impossible. Only the assiduous clamour created by repetition can transmit to us what only happened once.[40]

The "assiduous clamour" conveys visions inability to construct the temporal de-stabilization on which reading is based. One of the outcomes of this study is to suggest that Foucault's incorporation of hearing within his manifestly "objectivized" readings of power leave us with a sense of ambiguity about the relationship between authority and the "voices" (interestingly the word *la voix* in French means both "voice" and "vote") of the sonorous present.

## Modern France: The humanity of hearing

The way in which hearing might have acted as a catalyst for the construction of "humanity", that broader spatial realm of compassion and social cohesion which shaped social practice within eighteenth- and nineteenth-century French society, is a key theme of this book. French philosophical writings of the late seventeenth century demonstrated that hearing was not simply a single mode of "understanding", but was the most fundamental way to engage with a broader ethical and moral realm, a space of "humanity", through attention to sound and voice. In addition, medicine and science emerged as having all the necessary tools for the connection between hearing and spaces of "humanity" to be realized and represented. In Chapter 1 I demonstrate that Duverney presented human hearing as a universalizing resource within the human self by showing the complex way in which its material structure might make sound resonate. Claude Perrault (1613–1688) similarly conceived of a profound auditory world where humans might be connected to each other and material things through sound.[41] The consolidation of the relationship between hearing and place (in Nancy's "space-time" resonant sense) within the human ear also occurred in certain institutional domains. During the eighteenth century, certain key hearing cultures from the past were protected in the French urban environment because they projected more universalizing social ideals of inclusion embedded in their resonant hearing ethic. The authority of the Juge-Auditeur, (literally "Judge-Listener/Hearer") who presided over the small claims court at the Châtelet, as I show in Chapter 2, became one of the important figures in France during the eighteenth century because he offered a particular kind of auditory imaginative model fostering social coexistence with everyday people. In Chapter 3, I demonstrate that sound (the patient's cry and the sounds of the patient's body) provided the materials for carving out resonant spaces of healing which consolidate the doctor-patient relationship within the modern clinic. Finally, in chapters 4 and 5, I draw attention to the way in which the individual human body (firstly amongst the blind as a group, then the "everyday citizen") in association with technologies, became a site of the auditory imagination so as to connect with the broader body-politic.

To acknowledge such hearing sites is to re-evaluate Foucauldian ideas surrounding the development of "humanity" within eighteenth- and nineteenth-century France through a more nuanced sensory discourse encompassing the French cultural approach towards sound. Foucault's critical discussion of "humanity", a model of behaviour in modern history that encouraged greater social reform, has traditionally been interpreted solely through a simplistic concept of power that subverts ethics and compassion for domination and repression. This is despite a number of critical writings suggesting the Foucault's intentions were more complex. As Randall McGowen has demonstrated in his discussion of the anti-slavery movement, Foucault's model of "humanity" did ultimately recognize that "[t]he humanitarians recognized suffering and were sensitive to it. They knew how to give it voice and act upon it. The plight of the slaves, like the suffering of animals, was inarticulate until understood by the deep sympathy of the humanitarian spirit that translated and interpreted it."[42] Hearing becomes a central form of sensory activity within the humanitarian sensibility because hearing opened up a realm of being that was open enough to allow the repressed a voice, even if that "voice" was trapped in a discourse of power and domination. Foucault is careful to acknowledge the subtlety of this relationship into his description of the modern medical doctor who he describes as evolving from the figure of the doctor as listener and interpreter.[43] Foucault's depiction of the doctor's power here relies on a construction of authority that is imaginative enough to make the spatial slippage between individual and Other. This model became central to the French citizen's perception of a successfully functioning system of authority and to positive ideals surrounding a cohesive body-politic.

# 1
# Medicine, Science and the Auditory Imagination

In the past, to hear was to undertake an important personal challenge. *"Faith cometh by hearing"*,[1] quoted Blaise Pascal famously in 1656. Such comments were not simply theological statements, however. Rather, they were the first step in a much more formal and extended process of integrating the human individual into a larger constructed space of an ethical social experience. Philosophers who commented on the meaning of human hearing were attempting to find practical ways of improving society. Individuals had the capacity to improve their role within the social setting by tuning their ears. By the late seventeenth century, French anatomists and scientists began to think about the human ear in entirely new ways. Scientists such as Claude Perrault described how the individual was caught in a complex relationship with sound by describing the environment solely in terms of sonic objects. The anatomist Joseph Duverney identified the physiological structure of the ear as a site transcending expected boundaries of space and time.

Human audition became an intensely debated domain of research amongst a wider circle of different medical and scientific practitioners. By the late seventeenth century, key developments had been made by Galileo and by Mersenne on the theory of sympathetic resonance, which also came to dominate the thinking of eighteenth-century French scientists and anatomists.[2] Such work was grounded in the imaginative idea of the relationship between the human self to a broader cosmic sphere of existence, an idea that had been present since the time of Pythagoras.[3] Eighteenth-century French anatomists and scientists embedded the magical meaning of sympathetic resonance in their presentation of the human (and natural) auditory landscape. This realm maintained its transformative character not through any obvious reference to divine Creation. Rather, it was embedded in the imaginative construction of the complex mechanical processes of hearing and the hearing world itself.

## The humane listening world

By the 1680s, Cartesian philosophy, with a few notable exceptions, had dominated discussion on hearing.[4] The human body was blind and required sound to give it an ethical sense of its place in the world. Nicolas Malebranche had powerfully declared in 1674: "In short, it is of the greatest importance to make good use of our freedom by always refraining from consenting to things and loving them until forced to do so by the powerful voice of the Author of Nature, which till now I have called the reproaches of reason and the remorse of conscience."[5] Malebranche, like Fénelon, identified listening as a critical skill for potential leaders. He began his *The Search after Truth* with a warning of the serious political consequences that might result from faulty listening. Great leaders such as Alexander the Great and Julius Caesar had made the wrong decisions, he explained, simply because they had poor listening skills. Their ears were not functioning properly. They were attuned to the wrong sounds. Malebranche recounts how Caesar was constantly distracted by the "tumultuous din made by the crowd of flatterers surrounding him".[6] On arriving at the Rubicon River, he was so haunted by these horrible sounds that he sparked the famous civil war that ultimately damaged his people. Alexander too had been distracted. This time, not by a crowd, but by the strange language of the Scythians, which drowned out the "voice of truth".[7] Alexander had certainly been using his ears, but he had not been listening properly. For Malebranche, bad listening was demonstrated by an inability to strain out superfluous noise. This inability to engage in appropriate hearing was closely connected to erroneous judgments. The consequence of poor listening was not only the personal humiliation of individual error. More dramatically, it meant moral misery for the people: it was as a result of such poor listening, Malebranche explained, that Caesar sacrificed his country's freedom for his own ambition.

Vision was a very different case. Like listening, it could mislead. Yet Malebranche explained that men often forgot that human vision had only a limited field of view and might therefore distort the size of objects in certain circumstances. This was of vital importance in assessing the truth of scientific evidence. Sound perception, however, was much more directly connected to the general state of the human condition. For Malebranche, words could either make or break people's sense of humanity. Good listening had the power to liberate them from the often-abusive implications of much public speech: "But if men would learn to listen and to answer well, conversations would not only be very agreeable, but even very useful: whereas when everyone tries to appear learned, we only succeed in becoming swell-headed and in disputing without understanding: charity is sometimes offended, and truth nearly never discovered."[8] Observation itself required a form of elevated listening that went beyond the immediate action of seeing: "To submit to the false appearance of truth is to enslave oneself against the will of God, but to submit in good faith to these secret reproaches of our reason that accompany the refusal to yield to evidence is to obey the voice of eternal truth that speaks to us inwardly."[9]

Malebranche even encouraged listening in darkness. Such a process required both patience and practice. It might cause a deep emotional and physical pain, remorse and reproach. But this was a sign that the listener had renounced immediate emotional responses for a more profound emotional state of being: "If only God spoke to us, and if we judged only according to what we heard, we could perhaps avail ourselves of the words of Christ: 'I judge according to what I hear and the judgment is just and true. But we have a body which speaks louder than God himself and this body never tells the truth...'"[10] Though it was difficult to cultivate, such listening was ultimately liberating, since it insured the listener against bodily entrapment. The body required soulful sounds to fully exist as a form of life. Internal noises were often the least appropriate for emotional stability. To listen was to gain control over such sonic chaos. It was the ability to distinguish "the cacophony with which the body fills the imagination from the pure voice of truth that speaks to the mind".[11] Malebranche believed that with time and practice the listener was able to suspend judgment throughout their daily life and, in so doing, attain a kind of freedom.

Such freedom, of course, only occurred within the prescribed natural order. If the divine Being were silent, liberation would never take place. Human listening was considered essential in creating a connective zone of communication between the individual and the outside world. Through effective listening, individuals could actively participate in constructing a social domain, which forced them to improve their thoughts and behaviour. This kind of listening etiquette, Malebranche explained, should permeate every human action, even the philosophical writings of his own: "We would be very unjust and vain, then, to wish to be listened to like doctors or masters."[12] Other philosophers also drew attention to proper listening as a critical aspect of improving the human condition. Pascal was even more furious than Malebranche at the general state of social listening and also warned of its consequences for society as a whole. Poor hearing was again cited as a consistent characteristic of rulers:

> The mind of this supreme judge of the world is not so independent as to be impervious to whatever din may be going on nearby. It does not take a cannon's roar to arrest his thoughts: the noise of a weathercock or a pulley will do. Do not be surprised if his reasoning is not too sound at the moment, there is a fly buzzing round his ears; that is enough to render him incapable of giving good advice.[13]

Poor use of the ear was, for Pascal, a sign of man's folly. Kings surrounded themselves in loud and triumphant noises simply to scare people into submission. Intellectuals relied on a pompous tone of voice to over-promote reason. Parents gave their children bad advice in their career choices. Simple conversations were reduced to farce simply because people heard wrongly: "It is better to say nothing",[14] he advised, than to create the conditions for poor listening practice.

For Pascal, an elevated form of listening provided the perfect counterfoil to such pointlessness in life. It was the only act that might force the listener into an informed state of submission: "Know then, proud man, what a paradox you are to yourself. Be humble, impotent reason! Be silent, feeble nature! Learn that man infinitely transcends man, hear from your master your true condition which is unknown to you. Listen to God."[15] Proper listening stripped man of all his artificial power. It made man weak and taught him humility. For late seventeenth-century French philosophers and their scientific contemporaries alike, it was the natural, rather than the human, world of sound and voice that emerged as superior models of listening practice. Man might become beast through poor listening. Yet, more often than not, it was the sound world of nature that contained the sophistication and depth to match Enlightenment philosophical ideals.

Pascal advocated listening to sounds resonating around the natural world. One could only hear "all the conceivable immensity of nature"[16] if one listened properly. Man might remain fearful during this process but he was still able to separate the act of listening from the darkness of the unknown. Once he was really listening, he began to discover his own soul, and could only then begin to be happy. Fénelon also advocated listening to the natural world but referred to it as an intricate and harmonized system of social experience. His description of listening in his *Traité de l'Existence de Dieu* (1712 and 1718) is less concerned with the fear and noise of nature than the curious experience of nature's sonic language that drew man seamlessly into its web.[17] This approach led him to dispute the hierarchy outlined by Malebranche between the Creator and the natural world.[18] Nature, Fénelon believed, was the Creator. It supported man throughout his life, guiding him throughout his daily routine:

> The day is the time for social experience and work: night time shrouding the earth in its shades ends all weariness and soothes all troubles; it suspends; it calms everything; it spreads silence and sleep; while relaxing the body, it renews the spirit. Soon, day returns to call man back to work, and to reawaken all nature.[19]

Nature had its own spiritual voice.

Nature's voice, "a supreme and all-powerful voice",[20] belonged, according to Fénelon, to that of a higher Being. It transformed natural materials into wondrous things. But it could only be heard by man if he took the time to listen, he explained. If man truly listened to nature he would realize that intellectual reasoning alone could never make natural structures work. He would instead perceive "superior wisdom", a kind of white noise, a background spectrum of sound guiding natural beasts. This "instinct" enabled animals to defend themselves, sustain themselves and evolve.[21] Human bodies also contained such a noise. This hum ensured man's survival and protection. Humans could listen to the voice and direct their judgments:

> It is an interior master who makes me be silent, who makes me speak, who makes me believe, who makes me doubt, who makes me admit my mistakes or confirm my judgments: by listening to him, I learn; by listening to myself, I lose my way... The master who constantly teaches us, makes us all think in the same way. When we are quick to judge, without listening to his voice with self-distrust, we think and we talk of dreams full of extravagance.[22]

Proper listening was for Fénelon the perception of a universal voice permeating and uniting the globe: "While he is correcting me in France, he is correcting other men in China, Japan, Mexico and Peru, using the same principles."[23]

Humans listened, Fénelon explained, in order to survive their physical and social environment. The openness of the auditory conduit, Fénelon reminded his readers, was specifically designed for the preservation of life. Whilst the eyes were closed and at rest, the ears remained open, attentive to genuine warning signals.[24] The Epicurean doctor, Guillaume Lamy (1644–1683), also believed that nature provided the solution to an idealized socializing experience.[25] But he saw the human physical form as equal to nature in constructing this socializing discourse. In his *Discours Anatomiques* (1675) he went so far as to suggest that man "be respected by the other Bodies like the King of the Universe",[26] arguing that the Universe was not a hierarchical system of intelligence but a "proud strike"[27] which resounded through the very material of its physical parts. The human body itself was a series of interlinking acoustical forces, not a mechanical machine "moved" independently by a higher spirit. God or the "Author" was "a workman who did everything for himself, having no nobler purpose to consider. He produced the materials with movements without its different particles, and through this, all the bodies and an infinite number of unknown others that we see, have been formed."[28]

Through the human hearing process, natural systems and now even the anatomical parts of the human body itself, were revealed as sound-worlds of considerable sophistication. There was now little distinction between the effect of artificial and natural materials in the landscape when the proper listening process was called upon to make them sound in their very different ways. In 1680, Claude Perrault demonstrated in his treatise, "Du Bruit", that auditory physiology contained the material structure on which such sophisticated soundworlds depended: "I call Noise the effect of a particular agitation that meets in surrounding air and almost at the same time, in more distant air and as far as the ear."[29] Erlmann, in his brilliant analysis of Du Bruit, has drawn attention to Perrault's reliance on air *implantus* in order to make the connection between external sounds and the auditory system work.[30] For Perrault, listening occurred when external sound sources grafted themselves onto the mechanisms of the human ear, not in the form of a simple wave, but in mechanical meeting:

> The material in which the impression of the form of the sound is made. This material consists of two types of parts: the first type involves the dilated nerves mixed with a substance proper and particular to each sense organ: the other type includes those parts which are essential for the function of the closest organ.[31]

Erlmann explains: "Basically, what Perrault argued for was a qualita-
tive link between the physics of sounding bodies and the ear, the result
being an anatomy in which only those components mattered that maintained a specific physiological relation to the object perceived."[32] But
Perrault's work also confirmed in empirical terms that the soundworld
of the listening human being was highly sophisticated. The relationship between the ear and external sound sources was not reduced to
a simple formula. Rather, through a particular model, Perrault argued
for a much more accurate syntactical language to describe the complex soundworld, which philosophers had already identified, which the
human ear could perceive. The word *bruit* ("noise"), he explained, rather
than *son* ("sound"), was chosen because it acknowledged the variable
temporal qualities of particular sound-objects in the auditory environment. Sounds had many different lengths and qualities and this should
be formally acknowledged through an attempt at quantification, he
argued. "We do not ordinarily say the sound of a cannon, a caress, or
a mill, because these noises are not the kind referred to by the word
sound, which means a kind of noise which lasts longer than that of
the blow which produced it."[33] He was also frustrated by the way in
which the word *bruit* was used in combination with a number of vague
and descriptive terms. Perrault wanted to define such terms, *bruit de
choc*, *bruit de verbération* (impact noise, noise reverberation), for example, to demonstrate that the varied components of the soundworld of
the human ear were actually quite specific.

Perrault's work can be understood as an attempt to reinforce the
notion that the human subject could come into contact with an imaginative realm of experience by hearing. When the human ear came into
contact with the sounding world, the individual subject recognized the
world around them, but in a completely new way. What one thought
might happen did not happen. What one thought one heard, one
did not really hear. Perrault himself acknowledged the difficulty of the
task. Sound study required a certain ability to extend the imagination,
a "liberality" and a "magnificence of spirit",[34] he explained. Readers,
he warned, would confront paradoxes throughout. Perrault's theory of
particles demonstrated that auditory relationships were not uniform.
Whilst agitated air could make a tree move, it might not necessarily
affect the human ear and, by extension, "The agitation which makes
the noise usually touches the ear only and does not cause any real reaction in other more mobile bodies though it makes an impression on the
ear from a very long distance."[35] Hearing was, in this sense, very special.
Noise production also involved the absorption of other "air agitations"

which were contrary in motion. Because of compression, they had the capacity to join a number of different types of agitations into a single noise. These might be produced by different materials and emanate from different places. Agitations producing noise dispersed in every direction, not simply around an object or in front of it. Surprisingly, the agitation of air producing different noises was always of the same speed. Different objects had different numbers of particles and therefore different levels of power, yet the force of each one of these particles and their speed was always the same.[36] Perrault rejected the idea that air could be cut or sliced like a solid – a sound entering a flute was not cut in half by the tube but instead was fed into it – and threw into doubt the widespread theory that when air was cut it produced waves like a stone thrown into a lake. Waves such as these could only be produced using a flat surface and air was not flat. Instead, noise was also the product of agitated air hitting the ear in the same way as two bodies hit each other. It generated a level of turbulence that readily disrupted and diversified the simplistic notion of a single flat plane.[37]

Perrault was not simply attempting to define some basic points in modern acoustics: the role of density and elasticity in the propagation of sound. He was aiming at something much more ambitious. He wanted to define the human being solely in terms of an imaginative relationship with another world. The reader was now invited to conceive of the environment solely in terms of moving physical relationships, a place full of the tiniest air particles jolted into movement at great speed. This was the first time such a phenomenon had been scientifically represented. "Noise", for example, resembled the intense inflammation of a spot, for instance, on a piece of iron when it hits a stone. Perrault focused on two elements of noise production, the smallness of the space in which air is originally agitated, and the speed of the agitated movement. His model of noise production was a line of small balls. One ball at one end of these is pushed at great speed, causing all the balls to move. This model incorporated impact, compression and inertia. The smallness of the space in which these particles reside was not only contained within the two colliding objects.[38]

Perrault's study consists of a complicated list of sonic effects. Auditory objects glowed and glistened with the particular materials and with reflections of other objects. Sometimes materials of different types, balls of lead and balls of gold, for instance, made the same sound when thrown through the air. One might expect their noises to be different, yet, because of their movement through space, they sounded exactly the same. Perrault establishes that sound contains such a nuanced series of qualities that the listener always had to be attentive of the materials and physical actions at play.[39] Different noises, Perrault maintained, were classified according to their mode of action. The *bruit de choc* consisted of two solid bodies that hit each other and the *bruit de verbération*, a flexible body that hit another which was firmer. The *bruit de choc* was to be found in all types of noises, but that of the *bruit de verbération* was of a distinctive type being found in the voices of animals, flutes, wind, thunder etc. Each of these could be divided into simple noise, a single hit, or composite noises, (multiple hits). Composite noises had two different types, *bruit continu* (continuous noise), a kind of ringing (such as that produced by a bell or lute) from a single hit, or successive noise, *bruit successif* (successive noise), a noise which was produced by many different hits but that sounded like one, such as one which had "a voice which makes a long cry".[40] *Le bruit simple* (simple noise) had three different types depending on the actual atmosphere where the sound was created. A *bruit clair* (clear noise), for example, was a noise such as a voice extending unfettered into open space and *le bruit cas* (stifled noise) was a noise muffled, such as a voice behind a mask. *Le bruit aigu et tintant* (high-pitched and ringing noise) and *le bruit sourd* (deaf noise) depended on the nature of the bodies that caused them and the number of particles moving in the small space. Finally, the *bruit excessif* (excessive noise), such as thunder and the sound of artillery, were "excessive" because they were caused by the large number of particles, which were hit during their creation. The *bruit de verbération* might also be simple or composed (such as that produced by wind instruments). The simple were also divided into "small" (such as the noise of a whip produced by fewer particles) or "excessive", which dispersed particles on so many different materials, buildings, trees, rocks, water, thick clouds, that, "the ear is hit by them with quite an exceptional force".[41] Four types of air particle movements made up a continuous noise. A *bruit successif* (successive noise) had two different types, broken, *bruit rompu* (broken noise) and continuous, *bruit continu* (continuous noise), both of which tricked the ear into thinking it perceived a single sound. They could also be hard or soft depending on whether the hits were violently and frequently together or slight and separate to each other.[42]

Perrault's achievement, then, was to outline in empirical terms the soundworld described by contemporary philosophers. In this case, scientific experimentation was used to construct each aspect of the sonic experience as experienced through "intensive" and "proper" listening by the human subject. Technical language and experimental models in the study were critical in conveying both the materiality of the soundworld that the listener might confront but also its absolute complexity and sophistication. Perrault's insistence on detail and measurement can be interpreted as a methodology for the reinforcement of a philosophical point. That is, that hearing involved the construction of a complex imaginative world. Scientific language then became a useful tool in depicting the complexity of this world, which was now proven to have layer upon layer of textures, timbres and materials, and to ricochet and resound around the individual perceiver in a sophisticated way.

Later, in 1743, the Abbot Jean-Antoine Nollet (1700–1770) confirmed that the openings on the sides of fish, so-called *ouies*, were indeed listening ears. Nollet was very critical of scholars both ancient and modern who had presumed that fish were deaf. This was despite the discovery, he reminded readers, of certain key parts of auditory anatomy inside the fish body (ossicles, for instance) and known fishing practices employed by both Chinese and French fisherman using sound to lure fish to the surface. Nollet's work was less about the auditory physiology of fish, however, than about the implications of human audition in an auditory environment. He began to experiment on his own hearing capacity by plunging himself into the river Seine and subjecting himself to a variety of different sounds, a pocket gun, the human voice and whistles. Later, he set up a series of pneumatic experiments using water, pipes, thermometers and pressure gauges. Not only did he demonstrate that one could hear under water, but that sounds were not consistently dulled or silenced by water. Continuous sounds, for example, which passed from the air into the water, were better heard than those of short duration. Fish heard through the transmission of active sonorous bodies just as humans did, and this could be accurately measured. It was a different mathematical relationship, but the principle was the same:

> It could be that the vibrations of the sonorous bodies placed in the water communicating, as experiment has shown, through the parts of the actual water, make their impression on some part particularly designed to feel them and distinguish them; this part could have any anatomy besides that of the ear, and be placed anywhere else than at the place where this organ is found in terrestrial animals. In this way, fish can have a very keen sense of the noise and sounds which occur in the environment which is proper to them; it is therefore most important to them to know the various modifications.[43]

Soon after, Nollet incorporated his study of fish hearing into a larger study examining the relationship between air and sound.[44] Here, he demonstrated that most sounds were not simply the result of a single agitation but multiple ones, and that vocal sounds created by animals were just as mechanically measureable as those created by musical instruments and the human voice. He constructed a number of experiments to illustrate that the sonic world of the human was simply an extension of the natural. Wind was no longer a mystery. Even silence was now quantifiable according to auditory laws: "The vibrations of a sonorous body would occur in absolute silence", he explains,

> if there was not between it and us, some matter capable of receiving and transmitting this kind of movement: since the order of nature is such that a body does not act upon another if it does not touch it itself or through some interposed matter; of all those who have imagined exceptions to this general law, we cannot say that anyone has yet given proof of this.[45]

And, when man went onto the street or into the countryside, he understood why his voice could be heard better in the streets of the city than in the countryside, and why it might be heard even better than this in a closed room. It was for this reason, Nollet explained, that the deaf were always more unhappy than the blind. The blind could comprehend the world through sound. The natural and the artificial world now provided real evidence of audition's power to construct a powerful social domain.

## Hearing and the human ear

In 1677, Guillaume Lamy, prior to the publication of Pernaut's *Essais de physique*, completed his *Explication mécanique et physique des fonctions de l'âme sensitive* (Mechanical and Physical Explanation of the Functions of the Sensitive Soul).[46] In this treatise, Lamy outlined the different senses including hearing, the different passions that man feels on each occasion and the voluntary movements that result from such occasions. In contrast to the Cartesian view, he argued for a direct communicative link between the human anatomical system and objects of perception:

> It is not true as is thought in ordinary philosophy, that what we feel is in the object which provokes the feeling. The heat that the fire produces in us is not in itself, no more than the pain is in the needle which pricks us; but the needle is so formed that its point can stimulate what we call a sting or pain.[47]

After Perrault attacked Lamy for his "bizarre" Epicurean explanations of physical systems, Lamy published new editions of the *Explication mécanique* in 1681, 1683 and 1687, responding to Perrault's comments.[48] Lamy's response centred on Perrault's criticism of Epicureanism, however, he also expanded his work with a number of additions relating to the complexities of hearing research. There was a direct critique of "Du Bruit", an original physiological description of the ear, an original set of realist illustrations and a preface outlining a new discovery, supported by the surgeon, Jean Méry, of a field of nerves within the labyrinth. As a practising physician with an interest in Epicureanism, Lamy stood outside mainstream scientific and medical circles,[49] yet his substantial response to Perrault's work demonstrates the unique way in which hearing emerged as an independent form of medical and scientific research during this time.

Lamy capitalized on the hearing subject's deeper philosophical context in response to Perrault. With the additions and illustrations, the original text was converted into an analysis on hearing and its sophisticated domain of experience. Lamy explained that the study of auditory anatomy should never be complete since it related directly to judgment and morality, which were of central importance to the functioning of society: "Indeed, how do we know the interests of the Princes? How do we know how to conduct ourselves in matters of religion or how to judge trials for which we need ears, if we do not study the auditory anatomy?"[50] Thus, audition was presented as a unique area of study directly because of its philosophical relationship to the moral behaviour of the individual. Lamy went one step further in this auditory tirade by denouncing Perrault's use of the word *bruit*. By refusing to use the Latin-based word *son* from *sonus*, Perrault had demeaned any sense of human aspiration implicit in the listening experience. It did not follow that all sound must be noise,

> since in common speaking or in any science, we could not say that the sound of the voice, a lute, [or] a bell, is a noise. On the contrary, though it is true that in vulgar language, one does not say the sound of thunder [or the sound] of a coach, and so on, it is certain that in the language of philosophy, if one asks to what genre does the noise of a coach, of a door etcetera belong, one will quite rightly reply that it is a genre of quality that one calls sound almost in the same way that in vulgar language, one will not say that a man is an animal if one does not wish to insult him.[51]

For Lamy the issue of sound versus noise was not a small point. Sound acknowledged the presence of the "sensitive" soul in the human form, whereas the cruder term, *bruit*, did not. Sound, in his opinion, implied humanity, whereas bruit simply referred to raw, empty matter.

Joseph Duverney's much more famous 1683 treatise on the ear, *Traité de l'ouïe* (Treatise on the Ear) that appeared after this debate extended the philosophical premise of the listening subject to a new level. Perrault and Lamy had drawn attention to the tiny parts of the human ear, but their emphasis was squarely on the auditory scene outside it. Duverney's work was entirely focused on the physiology of the auditory structure of the ear itself. His use of anatomy, in this case, involved not simply the ordered display of auditory components but a striking sculptural re-representation of the ear's actual material landscape. There is no doubt that Duverney's work on the ear was inspired by the natural and cultural shapes of the Jardin du Roi, where he held the position of Anatomist and Chemist alongside the post of Doctor of the King. But it was also informed by his understanding of the body as an extremely fragile physical environment. Though he was a friend of Perrault and a fellow Academician, Duverney's interest lay not in the external physical world but internal bodily systems. Material throughout the bodily system, Duverney demonstrated, was so sensitive that it could easily become tainted by the tiniest of influences and was subject to blockage and decay. He published treatises on a wide range of medical topics ranging from apoplexy, child mutants, pancreatic problems and bone disease.[52] As one of the most complex and fragile systems in the human body, the ear was perfectly suited to Duverney's interests.

Duverney's work was closely indebted to the ancient Greek hearing model of Alcmaeon of Crotona (fifth century BC), who argued that hearing was not simply the penetration of the sound into the brain but that it arrived there by means of the ears, "because within them is an empty space and this space resounds".[53] This was contrasted with more simplistic temporal theories of hearing at the time (Anaxagoras, for example) who described hearing as simply a form of penetration of sound to the brain. Galileo and Mersenne completed major experiments on this theory by demonstrating the way in which different lengths of vibrating strings produced different pitches. Throughout the eighteenth century a number of different French mathematicians at the Académie des Sciences Joseph Sauveur, Daniel Bernouilli, Lagrange, d'Alembert and others perfected these theories culminating in Fourier's theorem in 1820.[54] Prior to Duverney's work, attempts to prove resonance theory by dissecting the human ear had been hindered by the extreme complexity of the ear as a physiological system. The Italian anatomist, Andreas Vesalius, Phillippus Ingrassia and Bartolomeo Eustachio, and later Antonio Maria Valsalva all made important contributions to acknowledging the different anatomical parts.[55] Yet the way in which the parts might interact with sound remained extremely unclear. When Duverney's work appeared, the Oxford medical professor Thomas Willis had already published a groundbreaking treatise on the ear, *De anima brutorum* (1672). Willis explained how sound was translated into nervous activity, which carry impressions of sound through the air in a wavelike motion from a sounding body to a tympanum. Penelope Gouk writes: "These impressions are then reinforced in the tympanic cavity and transmitted via the oval window to the inward air embedded in the cochlea's winding labyrinth, the cochlea being the true organ of hearing, where the auditory nerve endings are located."[56] Willis relied also on the idea of the animal spirits (which he also called the sensitive soul) to connect the hearing process to the faculty of imagination via the cerebellum.[57]

Duverney also advocated an appreciation of the ear as a space which shaped resonant audition through the complexities of its own material structure. However, he went much further than Willis in articulating key concepts of place theory, frequency-selective response, tonotopy and tonotopic projection to the brain.[58] These were dependent on his appropriation of the ear as a series of strange mechanical realms which created resonance as sound was processed. The treatise was inflected by Duverney's own awareness of audition's complexity writing, "The smallness and the delicateness of the parts of which it consists, closed in as they are behind other hard and almost impenetrable parts, makes finding them very difficult; their structure is so complicated that one has just as much trouble explaining it as one has discovering it."[59] Such an acknowledgment of complexity framed the way in which he presented the "objective" science of audition. It was not adequate to simply illustrate the different parts of the ear. Rather, it was necessary to break down the structure of the ear into clear transformative zones of resonance:

> It is not enough that some figures are true and faithful if they are not yet made and arranged in a way which removes all the ambiguity; I have shown the parts of the right ear still in their natural situation in order to conserve the first ideas that they suggest and to prevent them from getting cluttered and destroyed by others.[60]

These illustrations were combined with a written commentary, which described the cross-sections in fastidious detail.

The first set of illustrations focused on the ear's complex entrance. The ear might be considered in two parts, Duverney explained: inner and outer, hole and canal. But he then demonstrated that this was an absurd simplification. Even the outer part of the ear in itself was deceptive, demonstrating that it was not simply a "flap", but a sophisticated device wrapped in a double layer of material, the first, a layer of "thin and delicate skin covered with a little fat, particularly in young people", the second, "a nervous envelope" binding all together.[61] The ear, Duverney described in admiration, curls inwards like a shell, hence its name, *conque*, and it is never the same shape in two different people. Not only did it rely on a number of materials to create its form, it was powered by a further set of muscles, nerves and veins connecting it to the head. Two major muscles connect the ear to the skull, one descending downwards from a spot above it and one extending across behind. Arteries extended from the larynx and face, dividing at the jaw into two separate stems. These passed in front of and behind the ear, piercing the cartilage at different points, dividing again and again to cover both sides. Duverney demonstrated that it was no simple feat even to reach the end of the first structural stage of the ear, the tympanic membrane. On entering the ear, the reader was faced with an array of different shapes and textual surfaces. The conduit leading to the tympanum was not depicted as a straightforward structure. Part cartilage, part bone, it commenced in the shape of a tongue until interrupted by a number of openings. These were perilously connected by a special form of skin covering the inside of the conduit. The skin glistened with tiny yellow glands that resemble tiny organ pipes. They held thick, yellowy and gluey secretions of wax (Figure 1.1).

The next set of illustrations and commentary showed how the cartilaginous conduit led to the bony part of the canal, where it was confronted with a huge, strong ligament extending from the edge of the outer ear, along the cartilaginous membrane, to a small cavity in the temple bone. The bony canal protruded from the temple bone at such a strange angle that it appeared almost disconnected, suspended in space. Duverney identified the groove which held the membrane to the canal. But he was most interested in the texture of the membrane and what lay behind the membrane wall. The tympanic cavity, the *quaisse*, was surrounded by bone and closed in by the membrane. It was characterized by two conduits, the Eustachian tube, or *aqueduc*, leading to the nasal part of the pharynx, and a passage which connected with the sinuses. Duverney explained at length the combination of cartilaginous and bony tissues making up the tube and the "crossing" mechanism, which prevented air entering the ear. Duverney

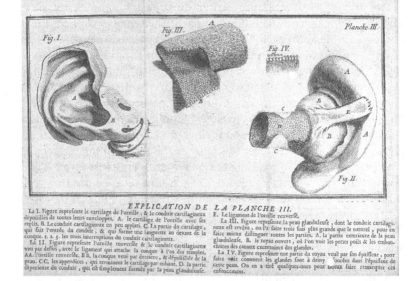

*Figure 1.1* Joseph Guichard Duverney, *Traité de l'organe de l'ouïe*, planche III (Paris: Estienne Michallet, 1683). BIU Santé, Paris

compared the ear at this point to the heart.[62] The ear, like the heart, he demonstrates, has a series of valves which prevent air from entering the mechanism from the wrong direction. Within the tympanic cavity the traveller arrived at a crossroads. The two openings or "windows" connected to a bizarre machine. One, holding the base of the *étrier* (stapes), a small bone and the other holding another membrane leading to the cochlea. Duverney described in minute detail the small bones transmitting sound to the cochlea. These were the *marteau*, (hammer), attached to the tympanic membrane with its niched head that connected to the *enclume* (incus). The *marteau* was attached to the tympanic membrane and the *enclume* to the *étrier*. Artisans described this mechanism as a hinge, explained Duverney. The two branches and the base of the *étrier* were like "a kind of chassis to the underneath of which a membrane is applied and glued just as oiled paper is applied to the chassis".[63] This membrane was delicate and "sprinkled with" blood vessels. He named a fourth *osselet* (ossicle) which connected with the *étrier* and the *enclume*, then two muscles attached to the *marteau* and a third attached to the *étrier*. This latter muscle is fat and fleshy at its centre and it created a strong, slender tendon that inserted itself at the head of the bone.

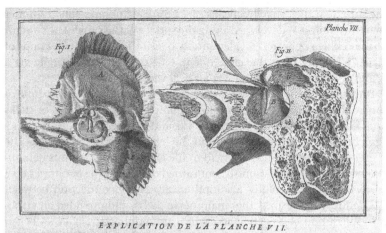

*Figure 1.2* Joseph Guichard Duverney, *Traité de l'organe de l'ouïe*, planche VII (Paris: Estienne Michallet, 1683). BIU Santé, Paris

Nerves might be confused with muscles in this crowded and compact place (Figure 1.2).

Entering the labyrinth was represented as if entering a maze. The traveller saw a series of openings of different shapes and sizes with extending ramps and windows. The vestibule of the labyrinth was dominated by the opening of the semi-circular canals, the upper ramp leading to the cochlea and the two branches of the auditory nerve. The oval window, leading from the *quasse du tambour* (drum box) was highlighted early on in his description. He described at length the way in which the three different parts of the semi-circular canals combined and curled. "The calibre of these canals is sometimes round and sometimes oval and it gets bigger towards the extremities like the bell of a trumpet",[64] he wrote. He emphasized the way in which the upper and lower semi-circular canals intertwined. The cochlea was encased in a bony canal which spun "two and a half turns around a core",[65] becoming thinner and smaller as it spirals inwards. The spiral lamina separated into two inside this canal and was attached at one end at the notch's base and at the other, on the surface of the canal, by means of a strong membrane, much thinner than the lamina. It developed along the surface of the canal. The conduit of the cochlea appeared divided into two as if there were "two ramps of stairs"[66] leading from the same notch. One entered the vestibule via a round window and the other via an oval window.

Duverney finally included a detailed description and illustration of the auditory nervous structure. It resembled a complex series of stalks and leaves which wind their way up and down the auditory structure, branching out to wrap around the head. Duverney confirmed the presence of air implantus (air held in the labyrinth), which was demonstrated by previous anatomists, but turned quickly to the auditory nerve correcting Thomas Willis' assumptions about the connection between the auditory nerve and other main nerve paths, and provided an elaborate description of how the nerve winds around the ear. Duverney began his description at the brain, describing the two branches of the nerve, hard and soft. The soft part of the nerve was made of the softest material of all the nerves in the lower stalk-like centre of the brain, with the exception of the nasal nerve. The hard part of the nerve was "so-called" because it is more fibrous and more compact than the soft part, but also because it led out of the skull before terminating, rather than being lost inside the auditory mechanism. After the two parts have separated, the soft portion divided into three parts, the largest at the base of the cochlea notch where it appeared to end. Duverney was careful, however, to note the presence of spiral ganglia, "several nets which are distributed at every step of the spiral lamina".[67] Again, he compared these auditory nerves to those in the nose. Hearing was as sensitive, in this respect he explained, as smelling. The auditory nerve was described in elaborate detail. It looped like the movement of an artist's pen. The two other branches of the soft nerve led to the vestibule connecting up again with the hard portion of the nerve, where both divided and created hoops inside the vestibule. The hard part of the nerve then led from a hole at the entrance of the petrous portion of the temporal bone. The nerve then led obliquely towards the tympanic cavity but did not enter it. It slid down this portion of the bone, descending towards the oval window, underneath the muscles of the *étrier* bone. It eventually left this bone via a hole between the apophyse mastoid bone and the styloid, connecting up with another major nerve path, the *cinquième paire*, then passed behind the ear and branched out. Nerves might intercept with mechanisms but they must not be confused with muscle material. Duverney drew attention to a small nerve, mentioned much earlier in the treatise, that is thought to have been a cord communicating with the tambour, but that simply ended in the tympanic membrane with no further nerve ending or smaller muscles connected to it. It was the optic, rather than the olfactory nerve that connected with the one running up behind the outer ear. This ran under the skin along the mastoid muscle and parotid gland, then, near the ear, separated into three pathways, which ran behind and under it, splitting into tiny pathways into the cartilaginous conduit.

Throughout this section, Duverney emphasized that the ear was at once very fixed and highly elusive. Only some of its parts mature as the human being matures in age. Specific pieces remained fixed from birth to life. Duverney ended the treatise by drawing attention to the membranous state of the bony canal in the ear of a foetus and its short length. The tympanic membrane was surrounded by a ring, which disappears as the baby grows. It appeared to shore up the membrane but begins to merge seamlessly with the bony canal after the age of three or four. The bony part of the aqueduct is also membranous in the foetus. Whilst the foetus was in the uterus, the tambour was protected by mucous material that also eventually disappeared. Yet the ear also maintained a level of independence and autonomy. There were no size differences between the tiny bones, cochlea and the semi-circular canals of children and those of adults.

In Part II, Duverney describes function, referring back to the illustrations in the first part. He discusses the directional function of the external ear and the way in which noise might infringe on the internal mechanism at different angles. But Duverney also demonstrates here how the materials and textures of the human ear might construct sound. The outer ear also had enormous muscle potential which tightened or dilated according to the violence or weakness of the air vibrations. The shape of the conduit was designed to protect the tympanic membrane, but also to render the sonic impression stronger. Wax, which Duverney described as a kind of glue, was designed to prevent dirt and insects, but it could also render the listener deaf. Nerve pathways inside the ear were also highly sensitive to foreign objects. The tympanic membrane was not essential for hearing (hence some deaf people could hear if they placed an object in their teeth, listening through their middle and inner ear) but if pierced it might eventually weaken and cause deafness.

The intricacy and delicacy of auditory mechanism was a key feature of Duverney's explanation of how sound is processed. He used the image of two lutes positioned on a table to demonstrate how the ear's mechanisms respond to vibration.[68] This was not as simple as shaking a solid object. It involved a very fragile set of materials and characteristics. The skin of the tambour responded to different degrees of resonance. It also relaxed and dilated according to vibration, creating a variety of noises and different sounds. The Anciens were therefore wrong, Duverney showed, to assume that the solid bone of the inner ear might receive such vibrations. The covering was simply too solid. Instead, it was the mechanically linking bones of the inner ear that transmitted the vibrations of the tambour to the labyrinth. These were the table on which are posed the two separate lutes; if one was disconnected from the table the second lute would not sound. Materials in the ear, such as the muscle of the *étrier* and the compressed air of the tympanic membrane also affected the sound.

Connections in the ear were complex and, acknowledging this, Duverney believed, helped in understanding sound perception. Air was pushed inwards and outwards through the ear via a series of mechanical valves. Whilst impure air from the lungs could not enter it from the nasal part of the pharynx, air from the nasal passage could. The nasal passage facilitated hearing amongst the deaf, who held an instrument between their teeth making the temple bone and the small bones of the middle ear vibrate. The deaf could also hear if one stands over their head whilst speaking. Their skull would then vibrate, activating the sensitive mechanisms of the middle and inner ear. Duverney emphasized the precision of the fanfare of trumpets in the semi-circular canal, as well as the cochlea spiral lamina. The smaller the bell of the trumpet the higher the sound transmitted "because the vibrations of these small circles are faster and more frequent".[69] Voice and ear were connected by hard portion of the auditory nerve with branches of the *cinquième paire* leading to the neck region. This is why men and birds sing and those who are born deaf cannot speak. Connecting nerves also caused the body to turn towards a sound and directly affected the pulse and respiration, hence the web of nerves that extended from the ear over the head (Figure 1.3).

Damage and disruption were disastrous to the fragile structure. The ear was incredibly irritable and sensitive to contamination. In Part III Duverney took the parts of the ear as he interpreted them one by one explaining the symptoms and signs of disorder in each: "I will not adhere to the divisions that Authors do ordinarily, but I will follow here the order of my description as I have done in the usages explanation."[70] He emphasized the aggressive character of auditory pain, beginning with its impact on the external part of the ear including the pinna, the conduit and the conduit leading to the tympanic membrane. He did not seek to explain the nature of the cause of the pain (though the irregular movement of animal spirits might cause referred pain from

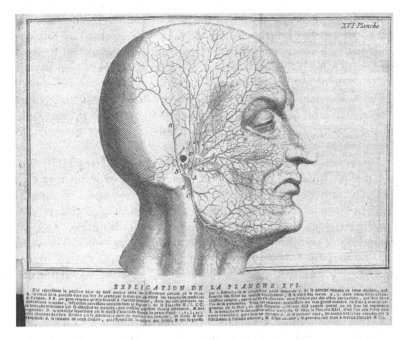

*Figure 1.3* Joseph Guichard Duverney, *Traité de l'organe de l'ouïe*, planche XVI (Paris: Estienne Michallet, 1683). BIU Santé, Paris

one part of the ear to another) but rather examined specific causes in each auditory part. He identified contaminated wax as a key source. This substance had been overlooked by the Anciens, he explained. Wax contained saline juices, which might be released and become so agitated as to cause an illness. Cold and heat had an adverse effect on the material, making it viscous and irritable in the conduit. However, wax was not the only possible cause of crippling disorder. Sometimes, the glands might evacuate serous fluid, causing "an unpleasant sensation".[71] Fluids which agitated nerve fibres caused four types of sensations in the outer part of the ear: tapping, tension, heaviness and pulsation. Such sensations were always accompanied by a high fever, insomnia, convulsions and weakness. Auditory pain in this outer section of the ear was so intense that it often caused death. The outer ear was a highly sensitive instrument. The membrane covering the conduit was made of the same material as the lining of the stomach and intestines. It contained more nerve connections than any other part of the body, and was closely interconnected with other parts of the ear, transmitting disorder easily: "After that, should one be surprised that the pain from the conduit is so cruel and so violent."[72] Duverney recounted a case documented by Fabricus Hildanus, of a young girl who by chance allowed a glass ball to enter her left ear. Unable to free the object, she experienced great pain all over the left side of her head and she became numb down the left side of her body. Her pain continued, she became hot and cold and maintained a dry cough, and experienced epileptic fits. Her left arm became thin. All these symptoms ceased, Hildanus maintained, when she managed to remove the ball.

Certain symptoms could be treated. It was not simply enough to protect the ear from the cold or wind using a piece of warm bread, as had previously been advised. The ear needed to be flushed or pumped out. The Anciens recommended plugging up the ear with cotton wool but Duverney also recommended injections made with sugar, as well as various other liquids mixed with almond oil or chamomile. The main aim, he explained, should be to heat all the parts of the ear, opening the pores and the openings of the canals, enabling material inside to flow freely. Inflammation caused by abscesses or ulcers in the conduit were more difficult to treat, especially those towards the end of the bony canal. Fluids became contaminated inside the conduit and might cause blood vessels to rupture. The corpse of a man who died from apoplexy revealed a thick, stinking auditory fluid. Obstructions were also dramatic. Thickening wax might "petrify", causing incurable deafness. A woman complaining of pain in the ear had tiny stones in the conduit that mixed with the wax. This formed a cast around the entire cartilaginous and bony canal. Sun exposure, Duverney recommended, might help cure this problem.

Foreign objects must be identified before deciding on a plan of action. Objects could be divided into those that might move or roll around in the ear (peas) or those which are hard and solid (lead balls or nodules). Soft objects stuck in the cartilaginous conduit could be broken down and taken out, whilst objects further down the canal might need to be lifted out via an incision from behind the ear. Duverney recommended the "Tenacula", which was invented by Hildanus, and resembles a long set of fine tweezers, but he also advised relaxing the conduit with almond oil before using the instrument. Phantom membranes might also form around the conduit, causing deafness. One should try to clean the conduit before injecting the area with appropriate remedies, he explained, but care should be taken not to pierce the tympanic membrane. Glands might also swell, blocking the conduit. Fumigation via injection was, again, a possible cure. The tympanic membrane could also become too relaxed, too tense, hardened or ruptured. Humidity caused over-relaxation. A tense membrane made the slightest sound unbearable. Fevers and headaches were other related symptoms. A hardened membrane was common amongst the elderly, causing deafness. Rupture was extremely painful but a common consequence of pushing something inside the ear with too much force or blowing out with the nose and mouth blocked. This was incurable. The tympanic cavity might be affected by bone decay, though this was rare. Parts of the labyrinth might be subject to abscesses of the membrane. These were untreatable and caused long-lasting damage to hearing, as was damage or compression of the auditory nerve.

Duverney's work supported contemporary philosophical writings on hearing's ethical meaning within the secular world by demonstrating the way in which hearing as a physiological system resonated sympathetically and sensitively within a complex material space. Perrault's work on the human auditory scene and Nollet's demonstration of the auditory world of fish also presented the idea of audition as a transformative realm of experience similar to that of the divine harmonic universe. The transferral of the theory of sympathetic resonance to a material site both within and surrounding the secular self had important consequences for the way in which interiorized spaces in general could be used to construct collective, universalizing social experiences.

# 2
# The Juge-Auditeur and Hearing the People

The research of Perrault and Duverney demonstrated that hearing involved access to an imaginative, broader realm of existence. This model can be used to interrogate the Foucauldian "humanitarian" frame of state authority and to identify an alternative narrative of power that evolved during the eighteenth century in France, encompassing the auditory imagination. Such a narrative acknowledged the value placed on human "voices" within emerging contemporary political discourses of human rights.[1] Foucault himself acknowledged the "voice" of the liberal subject in his understanding of the relationship between authority and the citizen in the modern state.[2] The way in which the everyday person was able to gain a sense of the right to be "heard" and the means to develop these expectations, however, is much more complex.[3] There are, however, specific examples of linguistic realms where the public had a clear expectation of being "heard" by those in authority before open debates surrounding democracy and rights. These sites of liberal expression were critical spaces of the auditory imagination where hearing cultivated a strong sense of inclusivity, incorporating an extended form of processing of "voices" on the part of the authorial hearer. By tracing the history of one such site from the eighteenth century through to the eve of the revolution, it can be demonstrated that audition, as practised in this transformative sense remained valued and conserved throughout the eighteenth century in France and was an important catalyst for political change.

By the 1700s, the Châtelet had built up a reputation as the most important institution, where the king, represented by the judiciary, might hear a particular complaint made by one French subject against another, regardless of their place of residence. This was recognized throughout the kingdom from the fifteenth century onwards.[4] Officers of the Châtelet, however high-ranked, were ordered never to abuse the Droit Commun ("Common Rights") of the people to make a formal request for a hearing at the site. The Châtelet's judicial influence covered the area of the Provost of Paris and the Ile-de-France yet, at least until the revolution, it continued to be revered as one of the most important sites of inclusivity throughout the whole of France.[5] Within the Châtelet, the Juge-Auditeur had "heard" the complaints of the lowliest subjects in the kingdom since the institution's inception. It was in this lower courtroom that judicial hearings most frequently took place. The position of the Juge-Auditeur was, in many ways, distinct from others within the judiciary, fulfilling a very particular role throughout the period of the Ancien Régime. The *Chambre de l'auditeur*, the lower judicial (small claims) court at the Châtelet, has generally been considered a secondary feature of Enlightenment history. Higher courtrooms dealing with serious crimes (Chambre du Criminel) or high-value commercial and family financial claims (Présidial, Parc Civil, Chambre du Conseil) have been more thoroughly researched.[6] At the lower courtroom of the Juge-Auditeur, however, "ordinary" working people ("it is, for example, the master, the servant, the labourer or the poor tenant and owner of the house, the stranger and the inn-keeper"[7]) came to resolve their grievances, to discuss and to seek relief from their everyday problems. The financial sums involved were small, and were described in formal judicial terms as being "small and of modest interest".[8] The Juge-Auditeur was listed as one of small group of lawyers titled *gens du roi*, that is, those officials at the Châtelet acting with direct interests in the crown and in public ministry.[9] He therefore maintained a different type of status to the magistrates and prosecutors of the upper courts, simply because he had a more direct stake in the general public interest. His position involved a highly specialized practice of listening, collating, judgment and accumulating experience in a different range of issues. Luchaire explains: "The *auditeurs* had therefore three distinct functions: they were judges, investigators and administrators."[10] Because of the number and nature of the complaints processed in his courtroom, the Juge-Auditeur was required to act efficiently, spontaneously and decisively. He demonstrated a strong moral sense in judgment, but also a particular ability to process the effects of the nature of the complaints and the effect on those involved. Since the majority of such complaints involved urban disputes, the Juge-Auditeur needed to be sensitive to the atmosphere of the street whilst also maintaining regulatory control.

During the long period leading up to the French Revolution, the administrative responsibilities of the Juge-Auditeur increased. This was despite the original definition of his role as merely a magistrate in training.[11] The king's official *procureur* (prosecutor) noted that magistrates in the upper courts were not behaving appropriately when they passed judgment on small claims cases, forcing large sums of money onto losing parties with minimal resources. This was, they explained, to the "great detriment of our impoverished subjects".[12] It was decided that whilst the Juge-Auditeur was no magistrate, he was nevertheless the most qualified person to listen and pass judgment on small claims. His authority was formally recognized with the declaration stating: "The Auditeurs of the Châtelet of Paris will enjoy the Rights and Privileges to which they are accustomed, and as enjoyed by other counsellors of the said Châtelet."[13] Even though losing parties still maintained the right to appeal, the Juge-Auditeur was permitted to pass sentence for a specific level of claim stipulated as "up to the sum of twenty five pounds".[14] By the early eighteenth century the Juge-Auditeur was again reinvested with full Royal Privileges, doubled in number and authorized to continue hearing cases for claims set at 50 livres.[15] In 1785, just prior to the Châtelet's demise, the sum was increased to 90 livres.[16]

The Juge-Auditeur attained a unique kind of status and character simply through his contact with everyday people and their disputes. His courtroom had a long-established history dealing with a multitude of matters such as cleaning costs, furniture removal and small business transactions.[17] After each case was heard orally, the hearing was recorded in the minutes along with the Juge-Auditeur's final judgment. On 2 January 1770, for example, the Juge-Auditeur who was appointed at the time, M. Jacqutot, passed sentence on the defendant M. Grange, printer, forcing him to pay 40 pounds to the complainant, a M. le Prince, Master Tailor. Central to the recording process is the statement that the prosecutor, M. Chavignan was heard in full: "HEARD, the said M. Chavignan read his plea."[18] Though sentences were not final (throughout the seventeenth and eighteenth centuries appeals went to the upper courtroom of the Présidial), the Juge-Auditeur carved out a particular position simply by responding, in the first instance, to what was a very lively and vibrant culture of petty dispute. At times the courtroom began to gain a reputation inside the judicial system for "running itself" without concern for guidelines put in place to regulate its activity. There were concerns that the Juge-Auditeurs were passing sentences when, at that time, they were forbidden to do so. More serious were the numerous later attacks on them by members of the upper courts for taking multiple "costs and salaries" directly from parties prior to passing definitive sentences.[19] Yet, when the issue was debated and discussed, the powers of the Juge-Auditeur were always either upheld or augmented.

The Juge-Auditeur maintained a unique position within the Châtelet system. On the one hand, his position was constructed through his own self-contained, authorial presence in the courtroom as the king's listening ear. People came to him to have their complaints heard and resolved. Yet, on the other hand, his close contact with the needs of the people made him a much more outward-reaching presence on the Châtelet stage that we might, at first, think. Layers of written evidence characterized his courtroom. Formality was apparent in the rigidity of the templates used to record hearings. Yet informality also pervaded such documents, appearing in the numerous and varied nature of the complaints. Judgments were not simply the result of rule and regulation, moral and authority, rituals of the élite, but of maintaining connections with the complex world of the everyday populace beyond the confines of the Châtelet site.[20] The Juge-Auditeur was one of the few public officials situated on the "frontline" of everyday Parisian life and was compelled to act in their best interests. It was duly proposed that the position of the Juge-Auditeur might be combined with that of the Commissaire which, by the late seventeenth century, had been established at the Châtelet site in the Chambre de Police: "The Service of the King and the public interest expect the [Police] Commissioners to be capable of gaining the confidence and respect of the people."[21] With his outward-looking stance, the Auditeur was identified as the only possible official who might be able to form a bond with the populace.[22] He was not, then, only a person of authority and command, but someone who understood what was happening "on the ground". This was critical to the new modes of control in eighteenth-century police culture. As Arlette Farge has demonstrated, policing the populace did not only require demonstrations of authority or enforcements of punishment. It was also essential to gain the people's confidence and their respect.

> Judge or pacifier: these two terms offer a commissioner the choice between a large number of attitudes of which he is the sole master, after all is said and done. This is where the sources surprise and astonish, more or less upsetting the well-planned structures. The commissioner, the classic image of repression, suddenly takes on new traits.[23]

Yet, when the official Juge-Auditeur, Jean-Louis Picard, stepped into the judicial courtroom of the Châtelet in January 1791, he knew that all that was around him would never be the same again.[24] Picard, "Juge-Auditeur, rue Sainte-Croix-de la Bretonnerie,"[25] as he was listed in the Almanach Royal, the official guide to the Royal administration, had been appointed to the position in 1785, and had been, at that stage, hearing approximately 16 cases of personal complaint in the courtroom assigned to him each of the four days of the working week.[26] In August 1790, after a series of furious debates in the National Assembly over the corruption of the French judiciary, the entire jurisdiction of the Châtelet was officially suppressed.[27] In a matter of months, the building of the Grand-Châtelet would be subject to a demolition order.[28] In general, the Magistrates, ushers, lawyers and other officials employed at the Châtelet met the announcements with howls of protest and bitter complaint. They pleaded with the government to protect their positions.[29] Picard, however, also facing the end of his career, formulated a completely different response, instead arguing for a massive salary increase on the basis that he was heavily invested in the revolutionary movement of reform. He was not concerned with restoring the Châtelet in the same way as his colleagues, only with ensuring that his position as a Juge-Auditeur was appropriately remunerated in the future.[30] At the end of his letter of request he emphasized that, as a Juge-Auditeur, he was closely affiliated with the spirit of democratic change equating the position with the reconstruction of *la Patrie* ("Fatherland"). He wrote:

> He [the Juge-Auditeur] received only the amount that his post had cost him; he only obtained enough to pay back his creditors to whom he fully belonged. What he owed his friends who helped him, what he owed to a sense of honour proven through his attachment to the revolution and what dictated his position towards the restoration of the Fatherland.[31]

The ability of the Juge-Auditeur to negotiate political reform during this period suggests that his authentic role as "hearer" of the people could be powerfully maintained within a new structure of reform even as the Châtelet itself was destroyed. Indeed, the way in which the demise of the Châtelet was couched within a particular discourse of failing to "hear the people" suggests that it was not the Juge-Auditeur who was the ultimate target of reformers. Rather, it was the institution as a whole that had supposedly failed. This was a result of its own backward and self-serving practices, demeaning its proper authorial role as a hearer of the public "voice".

## The blocked ear of the Châtelet

Squeezed along the edge of the Seine, the Châtelet occupied one of the most crowded, congested and unhealthy sites on the Paris map. Hoffbauer writes that this less salubrious side to the Châtelet's grandeur was reported in Declaration du Roi in 1672.

> Having discovered the poor state that the Châtelet of Paris is in at present, and being moved by the misery suffered by the detainees both as regards the confined space and the humidity and darkness of the accommodation, and the infection and poor air in the cells; all of this causes lots of harmful illnesses to those who inhabit them for a time.[32]

Even at this early stage in its history, the Châtelet was seen to be in need of crucial reform and renovation. Plans from the period demonstrated a wide variety of options for reconstruction, from simple renovations and room additions to dramatic reconfigurations.[33] Robert de Cotte's proposal from 1685 united the two different parts of the Châtelet into a balanced classical piece.[34] Running down the centre of the building was a solemn entrance marked by four pairs of columns, facing the rue Saint-Denis au Nord. At the back of this room was a flight of stairs leading to the prisoner's courtyard and different assorted courtyards for prisoners. In the event, only minor alterations were made to the Châtelet in the late seventeenth century.[35] Some 100 years later, on Sunday 9 September 1792, the Gazette Nationale reported that the commune's attorney had recommended that the Châtelet be demolished: "This demolition, planned under the old regime, will have the advantage of opening up the rue de Saint-Denis, allowing the inhabitants to enjoy the view of the Pont-au-Change, and will make the air infected by the vapours from the Morgue, the Fish Market and the butchers' shops, cleaner."[36] The newspaper also published the brief tale of a trembling old man brought to trial at the Châtelet, commenting: "Amongst a host of interesting traits, or amidst the anger and vengeance of the people, one likes to find natural kindness and the sentiment of the eternal principles of morality and humanity."[37] The newspaper then reported that after a series of questions, the magistrate acquitted the man, turned to his assistants and said, "Innocent or guilty, he tells them, I believe that it would be unworthy of the people to soak their hands in the blood of this old man."[38] After this came a cry of relief from his devoted daughter, the writer explained. The daughter's cry was a heart-felt reaction to the magistrate's "humane" judgment. Both were expressions of impending liberation from the inhumanity of the Châtelet space.

There were similar expressions of relief after the demolition had taken place. The death of 214 prisoners in the building in 1792 had produced a collective sense of horror and the Châtelet's end was welcomed. L. Prudhomme, who witnessed the destruction of the part occidental, wrote in 1807: "It has given us pleasure to see these towers of despotism demolished. We examined with horror, the enormous dungeons that locked away a battalion of victims; fortunately everything has been filled in... the rue de Saint-Denis has a beneficial opening; in addition, the inhabitants no longer breathe bad air."[39] Even Hoffbauer, who was not old enough to have seen the Châtelet himself, presented the building through the Gothic-Romantic vision of the neighbouring Cemetery des Innocents: "The old mossy walls of the Châtelet concealed dying human beings with expectations of horrible torment. Simple accused persons who were subjected to torture because of inflexible judges like Rhadamanthe, groaned and crouched at the bottom of real tombs, from where cries of pain arose."[40] For these writers the Châtelet would always remain a place of death, disease and destruction.

But it was the idea of "inhumanity" that was employed by commentators most powerfully in the period leading up to its demise. "Inhumanity" was reflected less in the suffering of the prisoners than in the supposed rotten internal workings of the courts. It was Mercier who most colourfully painted the picture of an institution infected by corruption deep within its core: "If you have a place in a house which is dirty, dark, foul-smelling, messy, and full of rubbish, mice and rats will move in without fail. So it is that in the depravity and abominable chaos of our jurisprudence, we have seen the race of prosecutors and ushers appear."[41] For Mercier, it was the hundreds of lawyers and ushers who transformed the building from something which might be considered simply hazardous to a remnant of living evil: "Our jurisprudence is only a mass of enigmas taken at random from the works of some jurisconsultants from a foreign nation; and when the different customs and laws are deprived of clarity, do not be astonished by the monstrosities of the procedure."[42] A major part of the problem, as Mercier saw it, was the multiple levels of justice in the Châtelet. These made the system sluggish and expensive. But it is the description of the lawyers living "handsomely in the labyrinth of the procedure"[43] that most vividly conjures up Duverney's diseased ear opened out on the anatomical table. Like a bloated insect inside the ear or a marble collecting wax, the hundreds of lawyers and their papers prevented the smooth working of the system: "There are eight hundred prosecutors, both at the Châtelet and in the Parliament, not to mention the five hundred judicial officers and all those who live from ink flooding across stamp-impressed paper."[44]

The blocked and diseased ear is not directly referred to in Mercier's text. It is instead more carefully implied through references to the Châtelet's popular status as an institution of elite hearing practice. For Mercier, the most serious implication of the blockage was that "humanity" did not prevail where it should – within the Châtelet site. To accentuate this charge, Mercier made specific reference to the lower courtrooms, "these subaltern agents"[45] where the problem was greatest since, unlike in the upper courts, they involved genuinely needy people. Such groups had been dragged into the system, he explained, often through no fault of their own and have been exploited by judges who destroyed whatever might be left of their family finances.

Magistrates were quick to respond to such criticism. One retorted, "Parliament is the main source of all the abuse committed,"[46] recommending, nevertheless, that the judiciary might be made more transparent through the publication of a legal register. This would force lawyers to listen more readily: "Because the order is established and made public, the Prosecutors must know when the case for which they are responsible must be pleaded."[47] Moreover, the pressure of public scrutiny would help prevent long and expensive court cases: "Because, finally, the Justice system would cease to be an unknown and inconceivable labyrinth."[48] Such comments underestimated the enormous task that would lie ahead. Post-revolutionary governments would face the challenge of constructing an entirely new justice system from scratch. They constantly repeated that this could only be successful if it was based on the ideals of the republican citizen-state rather than the privilege and privacy of the monarchy. But how was this to take place?

We will find that many of the discussions over judicial review surrounded the particular status of the Juge-Auditeur. The figure was, in many ways, the most important element of the judiciary to consider during change. Reformation of the upper courts could rapidly take place by removing privileges and high salaries and installing democratic processes of election, but the Juge-Auditeur was paid very little, charging on a per-sentence basis. He maintained his status primarily through daily contact with the people, as we have seen, by "hearing" them. This was the very quality desired of the new judiciary by the governments of the post-revolutionary period: "Society owes everyone security, peace and justice; Citizens must therefore be able to readily complain, be satisfied that suspects are dealt with, but sentences given only on full convictions. Police carry out, without inquisition, human and public justice, mild but inevitable punishments, such is the system of free countries."[49] The Juge-Auditeur was ultimately to be reintegrated into the system through the creation of the position of Justice of the Peace. As we have seen, this figure was exactly the same as the figure at the Châtelet. However, he would be authorized to "hear to the people" on the basis of a democratic mandate not monarchical privilege, and would be decentralized away from a single institutional site.

## The end of the auditeur? Diffusion or renewal?

In April 1789, Picard was elected to the General Assembly du Tiers Etat, Quartier du Marais, District Blancs-Manteaux.[50] He had been an active President of the District of Blancs-Manteaux and a President of the Representatives of the Commune of Paris.[51] His political fervour led him to play a key role in events during the French Revolution.[52] In July 1789, he read out an impassioned note written by the French military guards, accusing their commanding officers of cowardice when they refused to lead the march in celebration of the new political era.[53] In this speech, the guards openly attacked their leaders for refusing to listen to the people. Instead, he explained, they had employed "the most insidious caresses and the boldest lies to prevent them from following the movement of their Heart which would lead them to carry out the duties of citizens for the defence of the fatherland".[54] Picard was also one of the hastily elected judges in the criminal trial of Foulon, and he also signed the ordinance for the demolition of the Bastille.[55] His position in the midst of revolutionary fervour might seem strange. Yet instead of alienating him from revolutionary ideals, his position as an Auditeur during the Ancien Régime actually helped him to play a leading role in revolutionary change. In 1789, Picard simply transferred his skills as a "hearer of the people" and a purveyor of humanity, from the Châtelet courtroom to the sites of revolutionary change and renewal, the Champs de Mars and the Hôtel de Ville. His speeches and actions, all executed on the basis of his own moral authority as a Juge-Auditeur, were simply extensions of the same application of the role. No other Châtelet magistrate would have been able to offer such qualities at this particular moment. At a time when flexible leadership was most desperately required, the Châtelet Juge-Auditeur stepped into the breach.

In 1795, the office of the Juge-Auditeur was formally replaced by the office of the Justice of the Peace.[56] There would be one Justice of the

Peace for each village or suburb of over 2000 people. For suburbs and villages containing over 8000 inhabitants, more would be elected as deemed fit. Justices of the Peace were democratically elected by "active citizens" from each area. They had to be eligible for public office, and at least 30 years of age. Four "notables" from the community were in charge of nominations. They supervised elections and the period of candidature, normally of two years duration. Essentially, the Justice of the Peace took the same role as the Juge-Auditeur, listening to local quarrels and disputes, and passing sentences for small claims. They were more flexibly positioned away from a centralized site and multiplied throughout local areas. Picard, in a letter to the National Assembly dated 1790, insisted that if the Juge-Auditeur's courtroom was to be replaced by a network of Justices of the Peace, checks and balances still needed to be maintained.[57] He reminded the Assembly that at the Châtelet, losing parties generally had the right to appeal in front of a second "hearer", usually in the upper court of the Présidial. The request for a second hearing, whilst rarely activated, served to prevent the occasional error made by the Auditeur,

> but believe that it is no less dangerous to give this right of sovereignty (since it is really a right of sovereignty to judge in a court of final instance) to judges of courts of first instance, however modest the application, and whatever precaution one takes, when the application is more important, either to increase the number, or to be more careful and discerning in the choice of judges.[58]

The courtroom "hearing" must, by its very nature, generate the occasional error, Picard explained. This was not because it was futile or fallible in some way. Rather, it was simply because when the courtroom Juge-Auditeur entered properly into the humane listening stance, he surrendered himself to an extremely fragile world. He wrote:

> The prevention, and the inattention which comes natural to man: I am not speaking of the passions which corrupt the soul, I am speaking only of clouds which offend the spirit; all want, all shout to weak and fragile humanity that it is possible that a first judgment is an error, if not an injustice.[59]

Hearing, Picard reminded his readers, was not simply a mechanical procedure incorporated into an institution. It involved the activation of what might be considered an essentially a fragile material "spirit" world.

This, as Duverney so dramatically illustrated, was inherent in the act of human listening itself.

Picard never engaged in nostalgia for the Auditeur's courtroom. He never mourned the Juge-Auditeur's passing. Instead, he argued that whatever reforms were made, whatever new posts were created with the aim of promoting new ideals of democratic truth and justice, ethical hearing models should always remain central to the procedure in the court of law. In 1808, a decree reinstalled *un corps d'auditeurs* ("a body of hearers/listeners") near the court of appeal and the tribunals.[60] These *auditeurs* were somewhat different in character to their lone high-status predecessors at the Châtelet although they fulfilled essentially the same role. And, long after the Châtelet's demise, debates surrounding the position of the Auditeur continued to resurface. In 1828, Dufey attacked them as superfluous and indulgent. The original Auditeurs of the Châtelet, he wrote, were nothing but overblown *petit clercs* ("lower clercs").[61] They had been given too much authority and were therefore figures of corruption. Such statements produced a flurry of rebuke.[62] Later commentators acknowledged that the *juges de paix* ("Justices of the Peace") had evolved directly from the Auditeur system.[63] What these later statements demonstrate is that the Auditeur continued to hold a place in the national psyche well beyond the eighteenth century and the revolutionary period. The Juge-Auditeur, a "humane" listener with a sense of everyday concerns and problems, would continue to maintain an important role in the construction and function of the modern French state.

The transition of the Juge-Auditeur from eighteenth-century Châtelet official to roving Justice of the Peace reflects the general desire for decentralization, a trend which later thinkers identified in the post-revolutionary political landscape. This was, as Alexis de Tocqueville pointed out, a necessary result of the revolutionary "convulsion" that replaced feudal institutions with a more uniform and simple political order, one based on social equality.[64] Tocqueville also suggested, however, that centralization, the most fundamental institution of the Ancien Régime, fuelled such a change because it relied on a mediating process between authority and people's needs to shore up its character. The bureaucratic officials who went out to do the government's work, he explained, were central to this process because they enforced regulation through communication with the people. Nowhere was this more evident than in the courtroom of the Auditeur, where people came to have their complaints "heard" in the presence of an individual who was trained to listen to them. The power of the Juge-Auditeur and in the reinvented Châtelet space lay in an imaginative model of a socially inclusive space (made real in the Juge-Auditeur's courtroom) that functioned through the act of resonant hearing. This site of imaginative transformation, which was also critical to Duverney's presentation of the auditory processing space, depended on its own material mechanisms to transform hierarchical relationships into a cohesive space of inclusion.

# 3
# Hearing and Spaces of Medical Care

In his 1963 book, *The Birth of the Clinic*, Michel Foucault outlined one of the most important theoretical concepts for understanding the construction of eighteenth-century medical culture. "The gaze" is the term used by the author to describe the way in which French doctors of the eighteenth century began to contemplate the bodies of their patients and the spaces of their hospitals and clinics. Foucault writes:

> Clinical experience – that opening up of the concrete individual, for the first time in Western history, to the language of rationality, that major event in the relationship of man to himself and of language to things – was soon taken as a simple, unconceptualized confrontation of a gaze and a face, or a glance and a silent body; a sort of contact prior to all discourse, free from the burdens of language by which two individuals are "trapped" in a common, but non-reciprocal situation.[1]

Thomas Osborne responded to Foucault's concept in 1994:

> So what of the gaze itself? What does Foucault mean by this term? In *The Archaeology of Knowledge* Foucault criticized his use of the term (*regard medical*) since it seemed to imply a subject of knowledge. This is indeed the sense in which it has been taken up by some of Foucault's own followers, notably Armstrong, for whom the gaze is a kind of founding subject of surveillance. But in fact, looking at the pages of the *Birth of the Clinic*, there seems to be little support for this approach in that Foucault's actual deployment of the term there seems to represent less the intentionality of the perception than a particular, historically substantive, style of perception. As such, the

gaze does not originate from a particular kind of subject but is itself rather the effect of a certain kind of discursive constellation.[2]

Osborne interprets "the gaze" to be symptomatic of a much more extended "epistemological labour"[3] that forced the invisible in a patient's diagnosis to become visible.

Historians of medicine have argued that the main problem with the construction of "the gaze" is that it does not acknowledge the ways in which the patient shapes medical culture even as a repressed subject. "The gaze" does appear to articulate an all-encompassing mode of authority that constructs the medical world entirely on its own terms. Foucault writes:

> [The gaze] would scan the entire hospital field, taking in and gathering together each of the singular events that occurred within it; and as it saw ever more and more clearly, it would be turned into speech that states and teaches; the truth, which events, in their repetitions and convergence, would outline under its gaze, would, by this same gaze, and in the same order, be reserved, in the form of teaching, to those who do not know and have not seen.[4]

Foucault develops this model from a number of powerful visual metaphors which he applies to both the new medical architectural landscape and the invention of the clinic in eighteenth-century France. He uses the term the great "Œil de la Misère," for example, to describe a new form of medical "gaze", a new form of "totalisation" or "political consciousness" in relation to health care. In the clinic, the cadaver becomes the obvious visual prop for "the gaze" and is the source of a diagnostic approach that positioned disease as an important "deviation in life". Yet in both of these forms of medical culture, Foucault's definition of vision extends far beyond what the human visual field can supply. Instead, Foucault constructs an idea of vision that maintains an energetic, dynamic power with the potential to centre the source of perception in an imaginative transformative field.

> What now constituted the unity of the medical gaze was not the circle of knowledge in which it was achieved but that open, infinite, moving totality, ceaselessly displaced and enriched by time, whose course it began but would never be able to stop – by this time a clinical recording of the infinite, variable set of events.[5]

Such a description does not indicate a visual experience in a purely objective sense. Rather, it describes a form of vision constructed by a much more complex field of time and space which transforms existing relationships.

Foucault is more specific when he defines the clinical experience as "representing a moment of balance between speech and spectacle. A precarious balance, for it rests on a formidable postulate: all that is *visible* is *expressible*, and that is *wholly visible* because it is *wholly expressible*."[6] For Foucault the modern clinical era is not simply about "the gaze" but more about the way in which hearing is used in medical culture to recode the patient's "voice" for diagnosis by the doctor. In actual fact, Foucault relies on a sensory model which draws much more on the process of hearing than vision to transform the patient's "voice" into an object for authorial treatment and analysis. This is explained much more clearly in his 1966 article *Message ou Bruit?* in which he presented modern medical culture not through the prism of "the gaze" but through the perception of hearing and interpreting bodily messages.[7] Messages and codes, typographies and systems which the doctor "listens and interprets",[8] become expressions of the patient's "voice". These are the primordial sounds of the body which have been devocalized and organized into sentences. This final stage in the history of diagnosis is preceded by a gradual distortion of the patient's authentic "voice" into a new "voice" through its organization into codes. Yet, despite the gradual devocalization of the patient, hearing is retained as a primary mode of authorial power. In this chapter, I argue that this later model of medical culture developed by Foucault can be further used to emphasize the important ways in which eighteenth-century doctors and health reformers relied on a patient's "voice" or "voices" to recode medical and bodily space. As will be demonstrated, hearing was a critical tool in modelling spaces of the hospital and the clinic in order to process or interpret the patient's "voice". This reliance on hearing did not only involve the "rational" objectification of the patient's sounds for diagnosis as within the famous auscultation method of Laennec.[9] Rather, it entailed the careful accommodation of the patient's "voice" as part of the authorial doctor–patient relationship for the purposes of forging new collective spaces of social care. Within his work, Foucault does not simply erase the possibility of a benevolent attachment between doctor and patient. Rather, through suggestions of hearing he reminds us that more pastoral forms of power, models of "humanity" acknowledging the patient's "voice", provides the foundation on which the authorial model of clinical diagnosis is ultimately based.

## Hearing and models of hospital space

The new hospital space of the eighteenth century was, according to the Foucauldian model of "the gaze", a site where disease became a mark of social exclusion. Medical data in the form of scientific fact and empirical reason marked out the patient's "voice" as socially deviant. Yet, those who advocated the principles of "humanity" also used the patient's "voice" as a vehicle for the imagination of such spaces. As I have explained, "humanity" involved the construction of new sites of institutional authority within the urban space. As we have seen in the case of the Châtelet, such sites depended on the incorporation of "voices" for the authorial model to work. Hospitals too were shaped by a close relationship between ideals of "humanity" and care of the public "voice". In this sense, hearing played a powerful role in providing models for narratives of social reform. It was not that hospitals were literally spaces of sound when they were conceived. Rather, it was that because they "heard" the ailing public, the patient's "voice" resounded there within a resonant site of care. In 1786, the abbot de Recaldé wrote a scathing attack on the state of French hospitals, recommending the transferal of vast city hospitals, the Hôpitaux Généraux, to more healthy environments in the countryside.[10] The article concluded with a 46-point recommendation plan covering issues as diverse as the correct ordering of surgical instruments, accounting procedures, and the moral treatment of a patient's personal belongings. For de Recaldé, hospital spaces were the most important social spaces because of their dependence on society's humanity. The inhumanity of hospitals at the time, he explained, could be heard directly in the sound of moaning patients who had been treated badly: "In a Reign which fills the Nation with happiness and glory, we are surely astonished to hear, from all parts, the complaints and the moans of the unfortunate who are patients in the hospitals," adding that "the most precious thing for mankind is probably a society which provides a refuge in which a sick and poor man can get well".[11] At the core of the problem, he explained, was the destruction of a genuine sense of compassion, "the most beautiful of all our sentiments: this innermost sentiment which binds us to our fellow human beings".[12] This was not, he went on, directly the result of the recent increase in "frivolity, luxury, debauchery",[13] but the way in which such behaviours had worsened over a long period of time. De Recaldé characterized the inhumanity surrounding corruption and the hospital space as a kind of amplified noise which could be heard emanating from the "personal interest"[14] of the rich who were no longer alarmed by the "unfortunate's cry".[15] Such was the prominence of this selfish sound in society that it had ultimately drowned out the genuine voice of the needy. It was the perception of *le cri* alone that could restore humanity to the hospital landscape and society as a whole.

De Recaldé's definition of the hospital space as a zone of heightened listening was based on his own moral expectations of the rich towards the poor, "the most unfortunate class".[16] If the rich did not listen to the poor by attending to their health, he explained, society as a whole would collapse.

> We have always acknowledged that the most unfortunate class [of society] was, however, the most precious in the State for culture, the arts, and the population: this is not a new idea.... We must not then allow the strongest part which, amongst us, bears the greatest burden and provides the most services to Society and to which we owe so much gratitude and commiseration, to waste away.[17]

It was not that the rich did not want to associate with hospitals. On the contrary, they wanted to be seen and identified within broader social circles as funding and supporting hospital spaces. However, the rich were not actually hearing those in need, de Recaldé warned. This could be demonstrated in their failure to maintain hospital spaces and their desire to turn hospitals into revenue-raising institutions solely for their own benefit: "The Officers' tables are very well served in all the Hospitals where they also have comfortable accommodation. In several of these Hospitals the Poor have only coarse bread, little or no meat at all, and bad vegetables."[18] This accusation was pointedly directed at the nuns of the Maisons de Charité who, Recaldé explained, were supposed to be living a life of frugality whilst attending to the needs of the poor. Yet it was equally applicable to hospital managers, who were also traditionally employed at the poor's service. De Recaldé was shocked at the disparity between those who were in charge of hospital spaces and those who depended on them as a means of survival: "Indeed we see that the Treasurers, the Public Money Collectors and the Administrators are leading a delightful life in the midst of the misery; everyday these people show the unfortunate at whose cost they live, the tremendous difference between their abundance and the distress of those in their care."[19]

De Recaldé's text made a general call to the "humane" empathetic ear of the reader: "My weak voice, stripped of eloquence, calls on you to help the unfortunate."[20] Such a call involved the need to recognize unjust treatment towards the poor within hospitals when it was witnessed. Charity must be genuine, not feigned, he explained. It must not be used as a tool to manipulate. There was a tendency amongst the religious to discriminate between the needy on the basis of their own personal opinion. More worryingly, there was a complete inability on the part of the religious to offer the poor any kind of hope or relief from suffering. Consequently, the emotional call to listen with humanity in de Recaldé's text was closely aligned with his detailed description of measures for reform, including such advice as: Bedding should be regularly cleaned, patients should not be converted, surgeons should visit incoming patients immediately. The intense emotional experience of hearing the needy's call was intimately bound up with a whole series of appropriate and ordered responses: reformer-hearer-interpreter.

De Recaldé's call for a space in society that might help the sick, poor man to *guérir* was the product of an extended emotional attachment to the realities of the sick, poor man which came from listening to his cry for help from within the decaying hospital space.[21] De Recaldé demonstrated in his document that the process of public health reform in France involved much more than a simple shift from a culture of assistance to utilitarian, productive bodily reform. It was the result of a much more empathetic process of listening in order to harness a transformative sensibility that was powerful enough to initiate massive change. The issue of health care reform, the recalibration of the Parisian space to mark the post-revolutionary site described by Dora Weiner as the "citizen-patient", depended on confronting the old system and the needs of the poor with listening ears in the years before.[22] Hence what we find in the key pre-revolutionary and revolutionary documents on health care reform is a somewhat paradoxical combination of the decaying old system alongside new, ordered procedures of reform.[23]

By far the most powerful symbol to emerge during the health reform transition, however, was the hospital of the Hôtel-Dieu, the most notoriously unhealthy hospital in France.[24] It had been at the centre of a whole series of futile attempts at reform since the middle of the eighteenth century (and had even survived a serious fire) but had remained stubbornly resistant to change. Conditions in the large building on both sides of the Seine were becoming worse and worse, until the Parisian surgeon, Jacques Tenon, entered the debate over its fate in 1788. Tenon's report, *Mémoires sur les hôpitaux de Paris* (which was eventually presented to the Académie des Sciences), is famous amongst scholars for its detailed description of the modern medicalized hospital space, which was a radical departure from the open-door practice of the late seventeenth-century model of the Hôpitaux Généraux.[25] On close reading, Tenon's conceptual approach demands even further consideration. Hospital reform is revealed as a more open-ended and nuanced philosophy which demands that new biomedical realities and codes of procedure must be adaptable to the needs of the population. Audition was a critical aspect of this model since it facilitated the imagining of a new spatial environment enveloping the patient. The patient's "voice" is embedded in Tenon's definition of a hospital as a universalized refuge of "humanity".[26] Tenon demanded that the humane aspect of the institution's character must always remain, since it was its great strength:

> One is admitted there at any time, regardless of age, sex, country or religion: those with a fever, the injured, those with contagious or non-contagious illnesses, the mentally-ill in need of treatment, and pregnant girls and women; it is therefore the Hospital of the needy and sick person, not only those from Paris and France but from the rest of the Universe.[27]

The problem, as Tenon saw it, was more to do with the proportion and distribution of need across the entire urban space: "there is no longer any proportion between the Town, its surrounding area and the infirmary where the poor are crowded with four and six patients to a bed".[28] These are definitely products of "the gaze", yet the process of obtaining them required modes of audition which alone could facilitate the dimensional breadth of spatial imagining. Hearing facilitated the construction of a socially inclusive sphere for health care needs to be ascertained and adequately assessed.

> What method had to be followed in this research for it to be conducted to useful purpose?... It was about man, and the sick man: his stature determines the length of the bed, the width of the wards; his stride which is not as long or as free as that of the healthy man decides the height of the steps, just as the length of the stretcher on which he is transported, determines the width of the Hospital stairs.[29]

Throughout the text, Tenon reiterated the central argument that the Hôtel-Dieu failed in its role as *une sanctuaire de humanité* ("a refuge of humanity"), which was stipulated at the outset of the report. Whilst appealing to the call of the poor and ill, as de Recaldé so often did, Tenon added much more complex spatial imagining to "interpret" the patient's call. The technical construction of the hospital site might also be viewed, therefore, not only through the autonomous unit of the deviating body which must be healed or cured by "the gaze" but also through a complex technical imaginative landscape that consoles and connects (from an overarching authorial position) through the medium of audition. Such a landscape is an expression, as de Recaldé describes it, of "political charity", a form of government that is not simply a mode of surveillance, but also a marker of compassion, which articulates the notion that the risk of human disease and human suffering can be alleviated, at least to a certain extent, by protective action.

This protective aspect of eighteenth-century medical reform was strikingly demonstrated in the startling array of plans for the new Hôtel-Dieu.[30] The most famous of these was the circular design presented to the Académie des Sciences in 1785, prior to the publication of Tenon's report by Bernard Poyet.[31] The Committee praised the design.

> The arrangement of the wards ending at the two circular galleries, which are for general communication, seemed to us to be fairly well understood. This arrangement is infinitely preferable to that of the present Hôtel-Dieu where the wards are interlinked, or to that of most hospitals where the wards are interconnected allowing the circulating air to carry whatever comes out of one ward, into another.[32]

Poyet, who was later to design the impressive façade of Roman columns of the newly constructed Assemblée Nationale, was convinced that circular spaces were not simply healthier because they facilitated the circulation of air, but because they engendered a sense of patriotic security amongst the public. In his ambitious architectural project for a Cirque National in 1792 (that was, amongst many other projects, never completed), he explained that protection was implicit in its interiorized circular design: "At the least, the citizen must find a shelter from bad weather in these gatherings; no physical discomfort must in any way disturb their pleasures and the explosion of their patriotic feelings."[33] Poyet clearly envisaged urban architecture as spaces of democratic freedom and revolutionary fervour. Yet what facilitated such behaviour was the way in which his structures harnessed the public "voice" by emulating the imaginative mechanics of resonant audition. These were environments cutting across social boundaries within society at large, acknowledging the "needs of the people" through a symbolic form of hearing. Poyet's hospital was to be positioned on the often-flooded Ile des Cygnes and medical authorities had to admit that it would have been a disaster even if the required (and highly excessive) amount of funding could have been found. Nevertheless, Academicians couched their analysis not in criticism but in encouragement. In the 1786 report they wrote:

> We will praise the author of the Memoire who pleaded the cause of humanity so well, and we say again that the project for this hospital deserves to be commended; that its arrangement is fully understood and fulfils its purpose in many respects; that this construction will be superior for hygiene, for the comfort of the sick and the simplicity of the service provided at Hôtel Dieu.[34]

In his report, Tenon was even more reverential about the project: "There has never been a subject more worthy of fixing the attention of a learned company; everything makes it commendable: its aim, the wish of the Sovereign, the eagerness of the public, and the merit that would result from overcoming the innumerable difficulties that it presents."[35]

The authorities eventually decided on a plan to redistribute patients across four new hospitals, though it was never put into practice.[36] On paper, it reflected Tenon's topographical approach to medical reform, remodelling patient need according to air, space and water. Central to the plan, however, was the maintenance of the existing Hôtel-Dieu site as a "Maison commune".[37] Here, urgent cases would be received and materials for all the hospitals would be stored. Though the hospital models of Tenon and Poyet were never put in place, they demonstrate the radical thinking behind hospital reform from this period. Their significance lies in the way in which patient care was, for the first time, framed around a constructed sense of way in which a material space might most successfully interpret the people's "voice".

## Hearing and the clinic

In 1756, the Montpellier doctor Théophile de Bordeu (1722–1776), published his work on pulse, *Recherches sur le pouls par rapport aux crises* (Research on the pulse in relation to attacks on health).[38] In *The Birth of the Clinic*, Foucault positions Bordeu's work within the pre-modern clinical world of audition which was made obsolete by Xavier Bichat's pathological coupling of disease and death. Xavier Bichat, the pathological anatomist and physiologist associated with the Paris medical school, had demonstrated by 1802 that the brain and body were separate systems of vital power. According to Bichat there were two independent nervous systems in the human body: animal life in the cerebrospinal system, and organic life in the vegetative.[39] He outlined the strict independence of the two systems, so that the brain and spinal cord and somatic nerves on the one hand, and the ganglia with their vegetative nerves and ganglia, on the other, functioned autonomously, each with its own nervous power. Bichat's work had a major impact on perception of the senses in the first half of the nineteenth century in France, with the sensualist philosophers and their critical associates such as Marie François Maine de Biran and Victor Cousin arguing, as a result of his work, that the human brain connected to the sensory organs required a separate regime of health.[40] Such philosophers believed that since sensibility was the source of unstable vital phenomena, exercising the mind would guard against the physical and mental dangers of the imagination. Yet Mathieu-François-Régis Buisson, who had worked closely alongside Bichat was deeply critical of his ideas. In his work, *De la division la plus naturelle des phénomènes physiologiques considerés chez l'homme*, Buisson argued that the division between *vie animale* (animal life) and *vie organique* (organic life) should be replaced by *vie active* (active life) and *vie nutritive* (nutritive life).[41] The *vie active*, instead of containing all the senses, focused only on touch and the use of sight and sound in tandem with exterior objects. The *vie nutritive* incorporated taste and smell which, as a reviewer of the work explained, "only act on the tiniest molecules of the bodies, are capable of appreciating the most intimate nature of them, and are used to guide their choice".[42] Buisson was keen to analyse the role of hearing within the anatomical system, something that Bichat appeared to consider secondary. Buisson's philosophy of hearing consisted on the one hand of passive reception of sonic waves (an awareness of sound without actual listening), and on the other a much more important *audition active* or *auscultation* (specific listening to sound).[43] A reviewer explained that this later form of hearing

> differs essentially from passive hearing in that the hearing exercise is an act of will and attention is paid to it, and it is this difference which is marked in ordinary language through the use made of the words *listening* and *hearing*. We cannot stop ourselves from hearing, but we only listen if we wish to. We listen in order to have a better understanding, and indeed we can sometimes listen without hearing, when, according to the popular expression, we lend an ear to recognize the sounds for which we thought we had received a first impression.[44]

Such a mode of sonic perception, Buisson believed, belonged to *vie active*, which also included gesture, the movement of the head, and the projection of the voice.

Jacalyn Duffin has brilliantly described how Buisson's work became central to René Laennec's famous discovery of mediate auscultation in the clinical setting.[45] Laennec encountered Buisson at the Paris medical school and reviewed his thesis with favour despite Buisson's controversial status amongst the medical establishment at the time. The use of the stethoscope from around 1816 ensured that Buisson's principles of active listening could be put to direct clinical usage. Laennec was able to build a diagnostic practice through associations between certain sounds in the body and the detection of organic lesions. Laennec's work revealed pulse to be less important in the diagnosis of disease (particularly of the heart and chest) than had been previously thought. Yet Bordeu's work was not, as Foucault suggests in *The Birth of the Clinic*, so radically in opposition to that of Corvisart and Laennec. Rather, Bordeu, like Laennec believed that the body had an ability to create its own auditory narrative, which could be classified by the physician. It was the physician's role to identify and interpret this unique feature by listening to different pulse types. Different combinations of pulse patterns were ultimately shown to produce internal and external physical affects.

Bordeu's text was a powerful response to a debate in mid eighteenth-century French medical circles regarding the relationship between pulse patterns and musical law. He believed that pulse patterns were too complex to be compared to music. Rather, the language of arterial flow required its own set of organizing principles. He writes:

> It is quite true that the natural movement of the pulse can be compared, in general and in passing, to harmonies which result from the well-proportioned mixture of several musical instruments; however, it will never be anything more than a comparison which has no other use than to create an understanding of what must be expressed.[46]

Bordeu believed that the body could articulate a much more nuanced range of pulse patterns than the laws of music could define. Musical laws were simply inadequate and new methods of description for pulse patterns required further investigation.

In 1747, François-Nicolas Marquet (1687–1759), a doctor in Nancy, published his *Nouvelle méthode facile et curieuse pour connaître le pouls par les notes de la musique* (*New, easy and curious method for diagnosing the pulse through musical notes*).[47] After his death in 1759, a second edition was produced, which included a section on melancholia by the doctor and botanist, who was also of Nancy, Pierre-Joseph Buchoz.[48] Marquet argued that the body was a kind of "hydraulic machine"[49] powered by the heart: "As long as the movement of the heart and arteries is regular, the human body remains in perfect health, but as soon as this movement is disturbed by an accident, this healthy state becomes impaired by an infinite number of illnesses."[50] For Marquet, bodily health depended entirely on rhythm and sound. Equality and consonance meant health; inequality and dissonance, ill health. In his work, Marquet divided pulse patterns into two different types. A healthy pulse, *le pouls naturel* (natural pulse) was one which contained a balance between solids and liquids. It could be placed against the traditional musical minuet and found to sound at the beginning of each measure exactly (with six silent quaver beats between each pulse sounding). The *pouls non-naturel* (non-natural pulse) occurred when liquids and solids were not properly proportioned. It had a different kind of energy to the natural pulse and could be *simple ou composé* (simple or compound). The simple pulses were divided into binary opposites, *grand ou petit* (large or small), depending on the thickness of the blood, *égal ou inégal* (equal or unequal), depending on obstructions in the blood vessels, *profond ou superficial* (profound or superficial), depending on the strength of circulation etc. There was also a category of pulse called *le pouls tendu ou élevé* (stretched or high pulse) from broadened arteries, and *le pouls mou* (soft pulse) blood with too much serous fluid. Finally, there were different categories of quickened and slowed pulses, *pouls vite, pouls plus vite, pouls très vite, pouls précipité* and *pouls lent* (fast pulse, faster pulse, very fast pulse, quickened pulse and slow pulse) depending on the state of the blood, as well as more complex pulses, *le pouls intercadant* (irregular pulse), air in the blood, *pouls convulsif* (jerky pulse), and free and uncoordinated blood. All the non-natural pulses can be found in combinations called *les pouls composés* ("compound pulses"). Marquet transcribed all these pulse types using musical notation (Figure 3.1). Here, he used the minuet as a template, overlaying the perceived pulse sounds over the form of the dance. The method allowed for the notation of different pitches as well as their different rhythms and lengths. Combinations of pulses resembled musical scores with different rhythms and pitches heavily juxtaposed. Marquet believed that this system would enable the physician to memorize pulse types more easily. By visualizing the sounds of the pulse, a better diagnosis could be made.

In this edition, Buchoz sanctioned Marquet's work, describing it as "very ingenious".[51] "One cannot deny that the method Marquet has

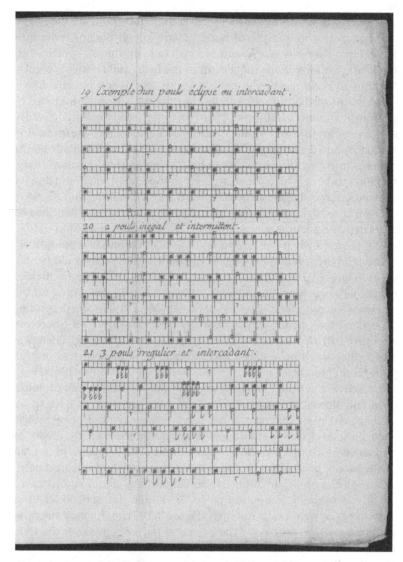

*Figure 3.1* François-Nicolas Marquet, *Nouvelle méthode facile et curieuse pour con-naître le pouls par les notes de la musique*, 2nd edn (Paris: Didot le Jeune, 1769), extract. Wellcome Library, London

given on the knowledge of the pulse through Music, is very ingenious and can serve to gain knowledge in this area of Symptomatology
so necessary in diagnostics and prognostics."[52] He then argued that
the relationship between "natural" music and the body was so similar that it could actually be used to heal the disease of melancholia.
This was a disease of the blood vessels which became smaller with fear
and sadness causing "weak and slow oscillations, and thickens the liquids through their stagnation; now, this thickening of the liquids gives
rise to melancholy".[53] Buchoz explained that it is necessary to "wake-
up" the nerves and help them regain "an equal and flexible tone".[54]
Medication would not help. A much better alternative was to listen
to music.

> Music whether vocal or instrumental, is either diatonic, the oldest of
> all, which ascends or descends in various tones, or it may be chro-
> matic which differs from diatonic only in that it is embellished with
> semitones; or, finally, it can be enharmonic, enriched by sharps and
> the softest inflections of the senses.[55]

Buchoz advocated active listening by the patient. Sound was considered
as an extremely powerful substance which was absorbed by the ear in a
sensitive manner. Oscillations from outside the body were able to affect
those inside. In support, he cited Duverney's work on the anatomical
structure of the ear.

> Sound is therefore propagated very quickly and reaches the ear
> instantly; then, through a wonderful mechanism, of which Mr
> Duverney has spoken so well, it hits the auditory nerve, by means
> of which it is carried to the *sensorium commune* and it is there that the
> idea of sound is formed.[56]

He then stated that hearing was similar to touch: "the organ of hearing
is a kind of [sense of] touch; the harsher the impact, the more it grates;
the softer the sound, the greater the feeling of pleasure",[57] explaining
that certain music would help cure certain conditions. "Dry" melancholia required music with a wide gradation of pitches whilst "Humid"
melancholia could be cured by fast and loud music. Simple music was
most helpful in restoring movement in the blood. It created agreeable
resonances in the ear, which then affect the body.

Bordeu's work was a direct attempt to refute the work of Marquet and
Buchoz on pulse. Musical laws were inadequate in explaining pulse.

It is in this sense only that one can say with Herophile that the movements of the pulse have a relationship with the laws of music; but if you were to apply the rules of music to the pulse, as a "Modern" has done, you would necessarily encounter difficult details, which would neither be helpful nor better-founded.[58]

Bordeu's more flexible approach towards pulse pattern was recognized and praised by doctors, who believed in the expressive and creative potential of the vitalist body. J. de Marque, editor of the edition, suggested in his preliminary remarks that Bordeu demonstrated the coldness of Mechanistic appraisals of the body (though he praised the "genius" of Descartes).[59] Mechanists had failed to understand and acknowledge the fluid nature of living systems. De Marque highlighted the issue of circulation which had fundamentally confounded recent mechanists. Circulation required new thinking.

I have so much to say about the famous circulatory system which has impressed so much, and which, for the Mechanics, has become an instrument that they have used with the same amount of liberality and confidence, as the Cartesians have put into the *matière subtile* [the idea of a fluid filling the universe]. And how this circulation has caused bad reasoning! How it has prevented Physicians from making good and frank observations on illnesses and on the "living" body, which formed the basis of ancient medicine![60]

Mechanists were portrayed as too rigid in their thinking. Ideally, doctors should think like a creative musician who wanted to comprehend the actual essence of musical creation: "I would really love it if a Musician, in order for me to walk in the footsteps of Rameau, taught me to preside at a concert formed by a large number of instruments, and enjoyed explaining to me, with an overly knowledgeable abundance..."[61] De Marque concluded that the musician must listen very carefully to diverse sonorities.

help me to grasp all of these different and various sounds, the combination of which makes the beautiful, the great, and the sublime of harmony; show my ear the means of grasping the slightest tone when it goes beyond its limits. Teach me the history of the living body, and I would say likewise to a Physiologist; we have analysed so much, and dissected so very much![62]

Bordeu's work was considered vastly superior to that of Marquet because it required the physician to listen more carefully, using his ear directly to form diagnosis rather than a notated musical score. In his 1767 *Nouveau traité du pouls*, Jean-Joseph Menuret, a Doctor at Montpellier, criticized Marquet's methods directly on such terms.[63] Marquet's work is described as "only an absurd and strange mixture of some dogmata of the Galenists, of those of the Mechanical approach, and the Chemists".[64] Musical notation might be useful in giving the physician an idea of the different pulse types, Menuret explained, yet, as a guide, it was far too abstracted from the body. Menuret continued: "His work would surely have been very advantageous, if the system, which forms the basis of it, had conformed less to that of the 'Mechanics', was less reasoned, and in a word, closer to observation."[65]

From the very beginning of his work, Bordeu demonstrated that problems in defining pulse related to the general difficulty of defining a complex auditory scene or musical work.[66] Pulse was presented as having a multitude of different auditory qualities which require careful streaming out. The oscillations of the body were not only difficult to describe in words, Bordeu explained. They required intense concentration to predict, often changed according to time and place, were difficult to memorize, and had so many different combinations that confusion inevitably resulted. According to Bordeu, the ancient Chinese and Galen both approached the problem by using imaginative descriptive metaphors.[67] These related to the rhythms and visual forms of the natural world and could be apparently universally applied: "He [Galien] claims to have found pulses which resembled the march of ants which he called *formicans*; others, which tapered off like the tail of a rat, he called *miures*; and, according to Herophile, he called those which he thought represented the leaps of a goat, pouls *caprizans*."[68] Physicians were attentive enough to acknowledge the complexities of pulse, yet in Bordeu's opinion their categories of classification were inadequate. He explained:

> Galien, in producing his Treatise on the pulse, reasoned much more than he had observed: he understood, however, that the different kinds of pulse had to be put into several classes: there was, however, some difficulty in categorizing them, in making them more recognizable, and even in expressing them in a fairly intelligible manner.[69]

Bordeu continued, explaining that more recent practitioners, the Modernes, responded to such descriptions by ridiculing them. He made lengthy reference to the early seventeenth-century Spanish physician, Solano de Luques, who along with other figures relied mainly on a more "direct" vocabulary of pulse description. Such a system was considered by the Modernes as much more effective than imaginative imagery because it did not distract the physician with unwanted and bizarre information: "The Modernes kept to simpler divisions and names, even more indicative in appearance: pulses were divided into *strong and weak, frequent and slow, large and small, hard and soft, etc.*"[70] Bordeu criticized such physicians as merely repeating the error of the Anciens. Just like the Anciens, the Modernes relied on abstract terms which, in themselves, meant nothing, he explained. Even if the physician relied on the new system, there were no fixed reference points for determining which descriptive term should be applied.

> But it is easy to recognize that this nomenclature adopted by the "Moderns" has almost as many faults as the one that they rejected, in fact these names do not express anything that is fairly accurate; it is not possible to determine what sign one must consider in the illnesses [to decide] that the pulse is for example, *hard or soft, large or small*; its *smallness* and its *greatness*, its *softness* and its *hardness* ... besides, it happens only too often that a pulse which is found to be *large or hard* by one Doctor, will appear *as small or soft* to another: consequently these definitions cannot express anything sufficiently positive.[71]

Though Bordeu chose to apply such descriptive terms directly to parts of the body, his motivation was fundamentally to define sound (specifically, the sounds of the body) in a more accurate way than his predecessors and contemporaries such as Marquet. He recognized that the auditory elements of pulse had, as yet, no fixed value system. Physicians continued to guess the meaning of pulse. The solution, Bordeu believed, would be to apply Galen's vocabulary directly to bodily parts and function (pectoral, nasal, guttural) and to physical bodily manifestations (critiques, non-critiques).

Bordeu explained that the different bodily categories used to classify pulse had three distinctive rhythms. The "inequality" or "equality" of pulse was the easiest sonic characteristic of pulse to recognize, he continued, but could be more carefully defined alongside an appreciation of overall bodily function. He then described how his use of functional distinctions (devised by anatomists) was the most detailed of all systems produced. Solano de Luques, he explained, never discussed the issue of "critical" and "non-critical" forms of pulse, nor pulses that related to human spitting, nor the issue of *pouls simples* (simple pulses), *pouls composés* (compound pulses) and *pouls compliqués* (complicated pulses). It should be noted that Bordeu's use of these terms differed radically from Marquet. Whilst for Marquet the *pouls simples* might have related to serious illness. For Bordeu, the term referred specifically to a period of stability within the illness itself.

> Illnesses in which crises are preceded and signalled by *simple pulses*, are never serious illnesses, unlike those presenting *complicated pulses* which are normally very serious: and yet it would require a lot for the various workings in the game of animal economy to be as noticeable and as recognizable in mediocre lesions of the functions as they are in major illness; it was therefore only in the exposition of the complicated pulses that examinations and discussions had to take place and this led to the productive and important rules that one was seeking to establish.[72]

De Luques also failed to make comments on the pulse of a person in a state of health. This, Bordeu explained, was essential in understanding the pulse of the ill. Overall, Bordeu emphasized the importance of detail in his own method, including exceptions to the rule. His aim was to push observations of pulse diagnosis "much further: & to bring them back, through this, to general principles which would enable as much light to be shed on the theory of art as on the practice".[73] Implicit in his desire to improve pulse recognition was the construction of a complex auditory soundscape relating to bodily illness. For Bordeu there was no division between pulse sound and medical illness. When the physician observed illness, he also perceived its sound.

Bordeu reiterated again the problem of abstract sound description and the need for an overarching system, when he begins to explain his method of pulse reading:"*The hardness, the softness, the greatness, the frequency* and so on, are only *states, relative modes* which can be assessed only through a common and fixed measure, to which one can refer all of these variations."[74] He recognized that pulse sounds are so varied that a more sophisticated measure was required. Before beginning an explanation of his system he reminded physicians that there were certain "fixed" sonorities associated with the nature of living systems that must be kept in mind. The first category of pulse explanation related to the age of the patient: "The natural pulse of the elderly is much *stronger*, much more *dilated*, and much *harder* than that of children."[75]

The second category related to the sex of the patient: "The natural pulse of women is, in general, brisker and closer to that of children…"[76] He then reiterated the importance of pulse rhythm and frequency (different from *la vitesse* (speed) and *la célérité* (celerity) in pulse, he explains). This could be measured almost exactly using a pendulum.[77] Establishing the natural pulse of the patient was also important so that irregularity could be diagnosed: "There are, for example, some pulses which will be called, *small, constrained, hard, full, dilated, developed*: it is as if one was saying that they are *smaller, fuller, softer, more developed*, than in the ordinary state of the natural pulse in the subject under examination."[78] Though all these features were significant, however, they only formed an incomplete picture of the true meaning of pulse: "Following the example of all Physicians, when assessing the state of a sick person, one should bring all of the symptoms together and consider all the circumstances: in how many difficult situations would we find ourselves in without taking this precaution?"[79] Bordeu emphasized the relevance of "fixed" guides to pulse. However, he advocates a much more flexible approach to pulse "observation" itself. This included attention to detail in assessing both the external and internal nature of the patient's demeanour.

The most important pulse category for Bordeu was "the natural and perfect pulse of adults".[80] It had a very distinctive sound described as "equal, its pulsations being perfectly alike and occurring at absolutely equal distances: it is soft, supple, free, not frequent, not slow, [and] vigorous, without appearing to make any effort".[81] Such elaborate descriptions also extended to other pulse categories. For Bordeu the most significant categories of pulse were the *pouls critique* (critical pulse) and *le pouls non-critique* (non-critical pulse). The non-critical pulse, which signalled nervousness or the beginning of an illness, had "too much sensibility, [and is] convulsive" whilst the critical pulse, preceding "the critical evacuations" was "dilated, developed, softened, extended".[82] This later type of pulse, Bordeu explained, was sometimes combined with "inequality". Critical pulse could be divided into two different types, *pouls supérieur* (upper pulse) and *pouls inférieur* (lower pulse), depending on whether the evacuation related to parts of the body above or below the diaphragm. The *pouls supérieur critique* (critical upper pulse) had three different types *le pouls pectoral* (the pectoral pulse), *le pouls guttural* (the guttural pulse) and *le pouls nazal* (the nasal pulse). These pulse types appear on their own, *le pouls simples*, together, *le pouls composés*, or combined with another type of pulse type, *le pouls compliqués*, such as a period of *le pouls non-critique* or *le pouls inférieur*. Bordeu then continued to describe the sounds of each of the different pulse combinations. The pulse of the *pectoral simple et bien declaré* (simple, and well-defined pectoral), for instance, was "soft, *full, and dilated with equal pulsations; one feels a kind of undulation in each one*; that is to say, the *artery dilates twice* but with ease, softness and gentle oscillation strength".[83] The pulse of the *pouls pectoral guttural* (gutteral pectoral pulse) was "*strong, with a repeat in each beat, and is less soft, less full, [and] often more frequent than the pectoral pulse; it appears to be intermediary between the pectoral pulse described in the previous Chapter and the nasal pulse which will be described in the following Chapter...*".[84] Bordeu finished this section by describing the rest of the pulse combinations along with case studies.

The range of sounds presented in Bordeu's pulse descriptions was extremely varied. Rhythm was important. However, words such as *fort* (strong), *mol* (soft), *ramolli* (softened), *etendu* (extended), *souplesse* (suppleness), *brusques* (brusque) and *plénitude* (fullness) referred to timbre as well as to the volume and length of pulse sounds. Rhythm was not simply a matter of temporal equality or inequality. Rather, it incorporated character, *avec une aisance* (with ease), and harmonic property, *dans chacune une espèce d'ondulation*, (in each one a kind of undulation). Bordeu's vocabulary extended beyond that employed by Marquet to specify the acoustical landscape. He attempted to describe in words particular properties of the pulse sound-object itself. Bordeu was careful to emphasize the importance of touch, *la finesse du tact* (the subtleness of touch),[85] in recognizing the different pulse types. Yet he also acknowledged that such a special kind of touch-knowledge relied on a memorization of pulse patterns: "The most clairvoyant and most confident Physicians with this sort of knowledge, are those able to memorize all the images of the different kinds of pulse."[86] The word *image* here referred not only to recognition of a pattern through touch, it also referred to the memorization of a sonic pulse pattern with all its acoustical nuances, as outlined by Bordeu in his different pulse descriptions. After connecting with the patient's pulse using touch, the physician then listened carefully, drew on his own aural memory and visual observation to associate each pulse type with a recognizable pulse category. This complex practice arose, Bordeu demonstrated, because of the highly creative and varied nature of different pulse patterns associated with the human system. Patients present themselves with many different pulse types. The physician then might construct an entire pulse pattern (often unique and original to the individual patient) using this special method of diagnosis. Patients sometimes presented only one particular pulse type, a pectoral pulse, for example, heralding the sputum which ultimately ends the illness.[87]

Others presented more complex physical and sonic symptoms, which required the use of two or three sonic "images" memorized by the physician. A patient, for example, might have a pectoral pulse, which became complicated with *le pouls d'irritation* (pulse of irritation).[88]

In Bordeu's work, the doctor-patient clinical relationship was framed around the ability of the doctor to hear and interpret the body's own set of "voices" which were not comparable to any external form of musical law. Bordeu's system relied on the body as a sounding structure to construct the clinical experience. Hearing acted as a mode of transformation in which the individual doctor became bound to the patient through a diagnostic processing of bodily "voices". Foucault's "gaze" from this perspective depends on the hearing subject just as much as visualizing modes of clinical analysis. It relies on hearing in order to carve out the transformative space of the clinic on which the pastoral or "humane" authorial relationship between doctor and patient was constructed. This powerful French cultural model of hearing which effectively combined the individual hearer and collective "voice" continued to shape institutions, technologies and cultural practices within late eighteenth-century and nineteenth-century French society.

# 4
# The Blind and
# the Communication-Object

In 1771, a group of blind musicians performed at the Saint Ovid's Fair in Paris.[1] The musicians had come from the Hospice des Quinze-Vingts, an institution housing the blind and partially sighted, situated on the rue Saint-Honoré in Paris. They were regular performers at the Café des Aveugles ("café of the blind") in the basement of the Palais Royal, and were often seen wandering the streets of Paris as individual performers or in bands. At this particular event, the blind were commissioned to dress as buffoons. An accompanying verse describes the scene: "It was lovely to hear these Blind people sing/And particularly nice to see them proud/Arguing as to who would give the best beat/To the songs that Paris flocked to listen to." In the audience was the translator and philanthropist, Valentin Haüy (1745–1822), who went on to establish the Institut des Jeunes Aveugles (Institute of Blind Youth), the eventual home of Louis Braille, some five years later. He was horrified at the event, writing some years later that when he saw the concert he knew instantly that the blind could do much better:

> Yes, I said to myself, seized with a noble enthusiasm, I will replace this ridiculous fable with truth. I will make the blind read. I will place in their hands volumes and instrumental parts that they have printed themselves. They will trace letters and read their own writing. I will even have them give harmonious concerts.[2]

Haüy opened his unusual school for the blind, the Institute for Blind Youth, in 1784, and was acknowledged, at that time alongside the famous Abbot Charles Michel de l'Epée, who invented deaf sign.

> Let us be permitted to pay homage to the talents and zeal of M. the Abbot de l'Epée who opened up the career of instruction to the Deaf

and Mute; following his example M. Haüy became the benefactor of the blind to whom this suffering part of humanity owe the means of happiness that one was unable to hope for.[3]

He then dedicated his time to the development of raised reading characters for the blind that eventually led to Braille, as well as specific educational programmes for young blind pupils.

Haüy's school and his treatise on blind education appeared just at the time when linguistic visions began to enter the ideological political arena. In 1791, Charles-Maurice de Talleyrand announced that a new form of French language would play a central role in developing the new French constitution. That language, as Sophia Rosenfeld has pointed out, "was the gestural and 'methodical' sign system developed in the preceding decades for the education of the deaf".[4] Haüy's work, though related to the deaf sign system was, however, much more squarely focused on sound rather than visual gesture. At the heart of his interest in developing sign systems for the blind was a belief that all citizens must engage in a more profound hearing process. When he saw the Quinze-Vingts musicians perform at the café, he was repelled by the sound of their music as much as the sight of the performance. By the time of the event, philosophical, scientific and artistic discussions about sound had reached their zenith. The composer and harmonist, Jean-Philippe Rameau (1683–1768), completed his highly influential and widely utilized, *Code de musique pratique* in 1761. This work outlined a comprehensive numerical musical system enabling instrumental musicians to create sophisticated improvisations in any form or genre.[5] This was the culmination of over 50 years of work on the science of music, including the publication of his *Traité de l'harmonie* (Treatise of Harmony) in 1722. Etienne de Condillac (1714–1780), in 1746, had insisted in his *Essai sur l'origine des connaissances humaines* (Essay on the Origin of Human Knowledge) that the development of modern language was dependent on the expansion of nuanced forms of sounds.[6] He explained that the *cri* in primitive societies was quickly modified by a whole variety of timbral and rhythmic characteristics, "violent inflexions", so that ideas might be more clearly and accurately expressed. From 1769, Haüy had worked as a freelance interpreter and code-breaker in fields as diverse as banking and the police, and in 1782, just prior to the opening of his school for young blind people, he was appointed official interpreter for the king, soon after becoming a member of Louis XVI's Bureau Académique d'Ecriture.[7] Haüy, like many other Enlightenment thinkers, understood music and language as sophisticated systems of sound. Reading a text or musical score or writing words or musical notation meant experiencing the phonetic sophistication of language. That is, this was as much an exercise in listening as lingual practice. Haüy was influenced by the work of his brother, René-Just (1743–1822), who became a member of the Académie des Sciences in 1783.[8] René-Just was a skilled acoustician and developed Nollet's theories on sound and propagation.

In December 1792, Haüy positioned his educational work with the blind squarely within contemporary revolutionary rhetoric. At a formal address to the 48 Sections of Paris, he reminded the "brothers and citizens" in the audience that the laws of Equality were most strongly applicable to his group of blind students.[9] But he did not simply demand equal rights for them. Haüy saw in his young blind – "my unfortunate children" as he called them – great potential for engendering a broader sense of social good amongst the people in the true republican spirit of "Human Rights". Already, he declared, "the Nation had elevated to a smiling Human Nature his sacred Asylum, the Institute of Blind Youth". This had facilitated the implementation of a "useful and consoling" education that had emerged from his own feelings of humane compassion, "my devotion to the care of Humanity".[10] Now, he demanded, all people should answer their call to come to visit them in their institution: "Blind Children invite you, through my mouth, to come."[11] There, all would experience his blind pupils' particular practice of crafting sound and be also able to fully participate in the culture of "humanity" that he believed only the blind as a collective group were able to create. Haüy invented communication-objects for the blind, technologies and practices that relied on systems of sound, directly for this purpose. It was the element of sound, whether experienced "live" acoustically or through the mediation of a touch-based technology, which enabled the listener to construct an inclusive space. Haüy's achievement lay in his determination to develop the idea that sound could inspire this state of "humanity" by shaping it into a material object which could be "heard". This notion, which came largely from his understanding of language as a mode of communication, was at the heart of his work.

## Listening and the art of expressive communication

Sound was contained within almost all the materials that Haüy offered to the blind. Gestural signs such as those used in contemporary modes of education for the deaf were obviously inappropriate for those who could not see. However, neither was Haüy interested in in oral dictation for the blind. He believed that they should construct sound into communication-objects, material things containing systems of sounds. Such objects were not modes of knowledge transmission, nor were they abstract symbols of a harmonized existence. Rather, they were practical interfaces, technological systems, which reminded the individuals who used them of the importance of collective social care by attending to an idealized listening state. Haüy's *Essai sur l'éducation des aveugles* (Essay on Blind Education) in 1786, emphasized that sensory sensitivities made the blind ideal candidates for developing and using such technologies.[12] Their innate abilities to feel sensation through touch and hearing were obvious. They had "natural dispositions"[13] for music and "seemed to become all Ears".[14] Their fingers were highly sensitive to textured surfaces: "Everyone is aware of the delicateness of this sense in individuals, who, from childhood, use it to replace that sense which nature has denied them."[15] The book itself was printed in raised characters so that the blind themselves could perceive it.[16] Conceptually, it resembled a "talking book", an object that speaks directly to the individual listener. Such new technologies were designed to draw the blind out of their individual state of isolation and into the care of a group. Listening was a central feature of these technical materials because it supposedly cultivated the "humane" qualities that facilitated such a process to take place.

Each letter of the alphabet was mounted on a metal plaque and a block of wood. These could then be printed as embossed text. The letters themselves were in cursive style so as to help the blind improve their handwriting technique that was also articulated in an embossed form. Such printing processes, both human and mechanical, were powerful constructions for the purposes of ethical hearing.[17] As the blind made contact with the letters with their fingers or pressing their iron nib deep into the extra thick paper, they were confronting a sonic system presented in material form. Haüy emphasized that the blind were not expected to become geniuses by using the system. He explained: "We do not claim to ever place the cleverest of our Blind in competition with any kind, even with the most mediocre of scholars or clear-sighted Artists."[18] They might not obtain regular employment, either, he declared, admitting, in addition, that their progress might be slow.[19] However, once the blind were exposed to the system, they could then be drawn into the "elements of the Sciences"[20] articulated by the technology. Their progress was not to be understood as a competition or even an "achievement". Rather, the blind were to engage in a continual process of attending to a higher hearing state. Music notation was also printed in a raised form. Music printed for the blind often contained emotional or spiritual messages which reflected Haüy's interest in hearing as affect: "When on your musette you sing with passion, a secret languor takes possession of my heart."[21] In practice, however, Haüy insisted on the reality of music as a sophisticated "Art".[22] He wanted the blind to learn instrumental music, Bach, Balbastre and Couperin, (to be situated in the Institution's Library), criticizing the raucous voices of blind street musicians, the Quinze-Vingts musicians, and their cabaret style of performance[23] (Figure 4.1).

Haüy's mechanical systems reflected his desire to maintain a structured, "humane" reality for the blind person. Technologies provided the level of sophistication that the ethical hearing experience, maintained through proper reading or writing or music-making, supposedly required. He explained: "Reading is the real means of embellishing the memory easily and quickly...Without it, literary productions would form nothing in the human mind but a muddled mass of vague notions."[24] In response to criticisms that it was pointless to have the blind read in this way, Haüy replied that reading cultivated a universal sense of kinship: "It's now our turn to question you," he wrote. "What use is it to print books for all the people who surround you? Do you read Chinese, Malabar, Jurchen, Quipu, Peruvian, and so many other languages which are so necessary to those with good hearing? Well! You would only be a blind person in China, on the banks of the Ganges, in the Ottoman Empire, in Peru."[25] Reading through the system was therefore a direct measure of situating the self through sophisticated, imaginative methods in a broader social realm.

## Communication-objects and the eighteenth-century public

Central to Haüy's plan was that the blind present their communication-objects to the listening public. Public presentation was an intrinsic part of their development and practice within the blind institution. Haüy emphasized that sighted audiences would be moved to compassion by them: "We finally call on you compassionate and generous spouses born into a comfortable existence; you whose son has just been born but will never see light; what sweet satisfaction for us to be able to ease your pain."[26] Haüy presented his young blind in venues as diverse as the Tuileries Palace, the Hôtel de Ville and Versailles, in front of audiences ranging from the Royal Academy of Sciences, to the Museum of Paris, to the king himself.[27] The blind did not only present musical and dramatic works they had learnt prior to the performance, but

*Figure 4.1* Épreuves des caractères des Aveugles, gravés et fondues pour être imprimés en relief, par Vafflard graveur-fondeur en Caractères d'Imprimerie, Cloître Notre-Dame, 7, inv. B-10-5003. Musée Valentin Haüy, Paris

also showcased themselves studying (reading, arithmetic) in front of the audience. These performances were not meant to impress or entertain audiences in an artistic sense. Rather, they were designed to forge a profound connection between performer and public audience through more sophisticated, constructed forms of speech. This was, of course, the central concern of rhetoricians who developed their practices throughout the late seventeenth and eighteenth centuries. Rosenfeld has drawn attention to a semiotic hierarchy established during the early part of this period in rhetorical treatises *La Rhétorique ou l'art de parler* (Rhetoric or the Art of Speaking) by the Oratorian, Bernard Lamy (1640–1715),[28] which pitched words above gestures: "Only words can ever effectively represent the mind's contents, since the laws of thought correspond to the grammar of words."[29] But it was a sonic, rather than a lingual issue that originally drove Lamy to establish his oral practice. Lamy had identified in the listening world around him a glitch in the communication process, a "discord".[30] Sound, he believed, needed reform just as much as language.[31]

*La Rhétorique ou l'art de parler* (1676, 1701), Lamy's most famous work, was the extension of an earlier work, *Nouvelles réflexions sur l'art poétique* (New Reflexions on the Art of Poetry) (1768), and it contained a lengthy criticism of profane poetry.[32] Modern poetry, Lamy explained in this earlier work, failed to attend to a higher, moral, hearing state. The ears of its listeners, for example, were deafened by pleasure: "Poets gradually studied how to compose their works according to the taste of their Listeners; pleasure was the only rule that they followed when producing their works."[33] Lamy criticized the contemporary poetic model of "the speaking painting",[34] a sonic product dependent on material beauty, for expressing itself without an inbuilt sense of man's fragility in the face of God: "In addition, Men do not see that God is the principle and the terms of this movement or of this inclination of their heart, which makes them love the grandeur and makes them seek blessedness in their present state."[35] Modern poets, he explained, hid imperfection, a necessary aspect of the true nature of the human condition and the essence of all creatures, behind grandeur. Excessive use of ornament (continually changing words, expressions rhythms), voyeuristic material ("chimeras, centaurs, monsters," "tales of heroes and captains," "the great sky and earth") and supposedly immoral narratives (appealing to concupiscence) all detracted from the proper employment of the passions and the inclination of the heart towards God.[36] Lamy's fear was that modern poets suppressed the act of higher listening amongst those who ought to be exposed to it.

Central to Lamy's discussion of practice was his desire to let a higher voice sound through the construction of material. The ugly promotion of "the beauty of the creatures" created a "discord" so loud that "the voice of nature which shouts constantly"[37] could no longer be heard beneath it. Modern listeners therefore allowed themselves to be dominated by the most superficial sonorities, "the playing of a bagatelle", he explained.[38] He demanded poets renounce exterior things and retreat inside themselves in order to let such a voice speak. Lamy considered his own poetic model much more positive than others because it expressed the poets' own faith in human nature. Modern poets were simply unable to draw more positive forces out of listeners because they themselves were unable to listen deeply.

> He is convinced of the vanity of creatures, and convinced that they cannot obtain this happiness for him through the forces that he desires: he knows also that he is unable to obtain this happiness through the forces that he finds in himself. He sees his weakness, but he does not seek the necessary help; he feels enveloped in dullness, but he does not ask for any torch to dispel it.[39]

Lamy did not deny that modern poetry contained useful things. But *la douceur*, that quality harnessed in verses by poets of ancient times, was missing from modern verse. This was the direct result of the failure of modern poets to listen properly. Audiences might be impressed by their work, but they were never fully engaged. Failing to listen, therefore, had serious consequences. It meant not only appealing to empty amusement amongst audiences. The more fruitful employment of the emotions emanating from the diverse passions (esteem, contempt, love or hatred) was prevented from attaining their true expression. Hearer and reader were forced towards a superficial sensation of happiness instead of experiencing profound feelings of emotion extending from their creator. Modern poetry, he admitted, contained a plethora of surface emotions. But these were squarely opposed to the most important religious passion: charity. Modern poems, in particular, also avoided the most criminal and dangerous passions, he explained. Yet in failing to attend to deeper movements in a listener's heart, they directly prevented its efficient release.

Lamy did not denounce modern practice for being too secular. Rather, he argued that it needed to be re-injected with meaning through attention to higher listening. The sensitivity of the human ear could be attended to even at the very basic level of speech construction, such as the choosing of specific words: "Tis True that Discourse is tedious when we give to everything that we desire to signify, particular terms, tis' tiresome to the hearer, if he has but common capacity,"[40] he explained. The ear could also be used to determine the correct placing of words: "Custom does not always observe Natural Order in certain words. It requires that some be placed first, and others follow at a distance. The Ear, being used to these kinds of array, perceives the least transgression, and is offended at it."[41] Tropes such as "Ironia", which relied heavily on the alteration of the tone of voice were entirely unsuccessful without taking the ear into account. One had to be able to perceive in the speaker's voice an important movement of an internal passion such as admiration, contempt, esteem and laughter towards a particular object, human, artificial or animal. Passions brought to life the soul of the words, "the Ideas present to our Mind".[42] Figures used to project the soul such as exclamation, doubt, communication and suspension relied on a more natural auditory posture. "Suspension", for example, depended on the fact that "the Hearer knows not what we mean, and the expectation of some great thing makes him attentive".[43] "Exclamation" required the hearer to submit to the "violent extension of the voice".[44] Lamy also explained that the passions had several different degrees of dramatic tension demanding several different auditory postures. Figures of measure, which required no resistance from the hearer, had a cadence, "agreeable to the Ear", whereas more turbulent, combative passions demanded not simply titillation but "the Judgment of our Auditors".[45] "We may, interrogate them, to fix and retain their Minds in more serious attention and make our reflexions upon what they have said."[46] Animating the ear and attending to a higher listening state was a central feature of mastering the profound art of speaking, declaiming and conversing.

Lamy was insistent about the importance of sound. In speaking, it was not enough to know the letters by name. Speakers should become familiar with their material sound.

> The Organs of the Ear are disposed in such sort, that they are offended by a pronunciation that grates upon the Organs of the Voice. A Discourse cannot be pleasant to the Hearer, that is not easiest to the Speaker, nor can it be easily Pronounced, unless it be heard with delight.[47]

In promoting auditory pleasure here, over pain, Lamy appeared to be contradicting his own aims of communicating deeper forms of emotion. Yet pleasure was compared to a deeper need for nourishment and self-preservation: "Prudence requires that we make use of this inclination to conduct us to our designed end; that we delight the Ears, which being the Porters of the Mind, may give our words more favourable admission."[48] Speakers are advised to balance consonants with vowels, making them sound softer and more fluent. Northern languages, English and Dutch should be avoided due to their inherent harshness. It was possible for a listener to become accustomed to such tones. However, in general, the state of one's spiritual level of listening reflected an inherent level of virtue "It is observed, that according to the different degrees of the peoples' inclination to delicacy, their words are composed of Letters more or less soft: they having less regard to follow reason, than to tickle the ears."[49] The number of thoughts in a sentence also affected the ear. If they were insufficient, the hearer remained frustrated. Too great a number of words and thoughts created confusion.[50]

Lamy also drew attention to the ear's timbral sensitivity, advising that because "sounds strike with moderation upon the Organs of Hearing. Those sounds which offend them are irksome and disagreeable".[51] Sounds should nevertheless be strong enough to make sufficient impact and be well proportioned in tone and speed. Quoting Cicero, he explained that, "the Ear is hard to be pleased. We many times displease, when we design to please them... Diversity is necessary, because the pain being divided, each part of the Organ is less oppressed."[52] It was advisable, he insisted, to combine diversity and equality in a sensitive manner. Curiosity, vivacity and judgment in a speech fashioned a solid and elaborate auditory imprint. It resounded long after the death of the author. Because men were blind it was necessary to attend to the deepest aspects of their auditory selves: "Orators acted by true zeal, are to study all possible ways of gaining their Auditors to the entertaining of truth."[53] In addition to creating beautiful sounds in this way, the speaker should adopt an air of honesty, civility and modesty: "Nothing so certainly alienates the minds of his Auditors and inflames them with sentiments of hatred and disdain, as with the vanity of self-applause."[54] Lamy was fully aware that he was proposing a discourse of artifice. Yet he also reminded his readers that it was grounded in a profound spiritual ethic: "It follows not but we may profess love to our Auditors, and insinuate into their affections, when our love is sincere, and we have no design but the interest and propagation of truth."[55] Lamy ended his treatise by exclaiming that the different parts of the discourse should be shaped according to the hearer's memory, desire and respect: "A Hearer is susceptible when he loves and listens to what we say."[56]

For Condillac, nuanced sounds in speech fashioned a society that was held together in a more robust bond of care. He believed that sound was a crucial force of social influence at the time when language was originally conceived. The civilized language of action was as much visual as sonic. Language evolved from "cries of each passion"[57] amongst the Anciens, he explained, which were enlarged through the action of gesture, but also through the shaping of certain sounds: "Now the natural cries necessarily introduce the use of violent inflexions; since different emotions are signified by the same sound varied in different tones. 'Ah', for instance, according to the different manner in which it is pronounced, expresses admiration, plain pleasure, sadness, joy, fear, dislike and almost all the passions."[58] Condillac even suggested that specific words were constructed according to the sounds of nature: "Finally, I might add that the first names of animals probably were made in imitation of their cries: a remark which is equally applicable to those that were given to winds, rivers, and to everything that makes a noise."[59] Condillac concluded that the sounds of early forms of language were so subtly expressed, that they resembled a kind of music. Thus, he argued that modern modes of declamatory speaking must incorporate a greater variety of sonic nuance in order to communicate their message effectively. Music and speech had lessened in their dramatic effect and become entirely separate modes of discourse, he explained, because they failed as essentially sonic modes of expression.

> The most perfect prosody is that whose harmony is best adapted to express all sorts of characters. Now there are three things concurring to harmony; the quality of the sounds, the intervals by which they succeed each other, and the movement. A language must therefore have sounds of different softness, even some that are rough, in a word, some of all kinds; secondly, it must have accents to determine the voice to rise and to fall; thirdly, by inequality of syllables, it must be capable of expressing all types of movements.[60]

Later works on rhetoric also emphasized the importance of sound in effective speaking. In his work, *Rhétorique française*, Jean-Baptiste-Louis Crevier wrote: "The ear is like the vestibule of the soul. If you hurt the ear through an unpleasant sound, the soul will be ill-disposed to receive what you present to it."[61] The speaker should not only have a vivid imagination but also a sensitive and delicate ear since the sound of words directly supported the speaker's thoughts. Students were guided by "good rules" and "excellent models" which they could study before fashioning their own examples.[62] Gabriel-Henri Gaillard directed his work on rhetoric towards women who, he argued, could then make a more profound contribution to society by learning the art of eloquence.[63] He reduced the principles of eloquence to three specific forms, *le Judiciaire, le Délibératif et le Démonstratif* (the judiciary, deliberative and the demonstrative), each of which were constructed according to the listeners in the audience.

> Indeed, the Orator, whatever subject he is dealing with, has to warn the Prosecutors and Judges to persuade; in the Object of his speech he has a client to be defended, a cause to be pleaded, a Proposition to be clearly established and strongly proven; finally, he has a vehement and fast summary statement to be made of his strongest evidence.[64]

Gaillard's focus on the "plea" in rhetoric was once again drawing attention to the humane listening ethic (one is reminded of the Juge-Auditeur at the Châtelet). The structured plea was a concrete way in which listeners could be made aware of the broader world of human need. As Gaillard explains: "In order to provide all of these grand effects, you have to start by pleasing: it is the powerful spring which moves the entire machine of the mind and the human heart."[65]

It was not simply within structures of rhetoric that sound could be manipulated. Rameau reduced some of the more complex acoustical theories of Cartesian mathematics into a single palatable system of music-making. Students could memorize the progressions, absorb them and practice them over a year or more. Only after this process of intensive learning and practice could they produce performances on their instrument of choice and compositions of real "taste". After memorization, progressions could be skilfully reversed to create a perfect arch form, and in turn, the invention of a complete piece of music. Rameau's system catered especially for blind pupils, a point he makes plain in the full title of his 1761 teaching manual, *Code de musique pratique, ou méthodes pour apprendre la Musique, même à des aveugles* (Practical Music Code, or Methods for Teaching Music, even to the Blind). Blind students could shape their fingers and tune their ears, producing the best results precisely because they could not see.

> Using the system, the fingers acquire a knowledge which, nourishes the ear of all the harmonic routes, it presents the mind with a reliable example of all the rules in which it must be educated with the result that judgment, the ear and the fingers of intelligence work together to quickly produce the perfection that is desirable in this genre.[66]

The system can best be described as a kind of physical intuition based on strict harmonic practice. It was directly constructed to demonstrate the lack of sophistication in street music.

> Listen to the people who sing and what they shout in the streets; nothing proves the pure effects of Nature better in such cases... from here comes music which is continually composed in a Tone which is varied only by that of its "fifth" as are Airs of Trumpets, Horn, Musette and Vielle [hurdy-gurdy]; these do not have any effect on the soul, unless it is through the variety of the movements.[67]

These sounds, Rameau explained, left listeners feeling cold because they relied on an overly primitive theory of harmonics. To create a successful communication-object, a level of sophistication was required.

Haüy's work with the blind in the late eighteenth century can be situated within this powerful cultural world of the communication-object. His own system of raised reading characters was, in the same way, simply another rhetorical form of presenting sound in a structured state. It was, then, no different to Lamy's poetry, Condillac's language, Cuvier and Gaillard's verbal rhetoric or Rameau's music. The aim of all these works was to fashion meaningful modes of communication through attention to sonic detail. Only by attending to the sensitive ear could poetry express charity, language, ideas, rhetoric, thoughts and music, taste and warmth. The aim of all these communication-objects was to demonstrate the importance of "humanity", an understanding of other people's needs by processing sound in a structured state. During the revolution, however, concern over false sounds, the very issue that had motivated Haüy himself to create his programme, began to surface amongst authorities in a dramatic way. When we trace the story of Haüy's career during and beyond the revolution, we can discern that whilst the ethical listening process still drove the public agenda, it was positioned within an entirely new atmosphere of fear and suspicion. Haüy became a target not only because his political motivations were considered dubious. His imaginative interest in sound ultimately exposed him to charges of deliberate subversion.

## The Terror and the problem of the *faux bruit*

In 1791, Haüy's Institute for the Young Blind was united with the school for the deaf established by Abbot de l'Epée at a site at the former convent of the Celestines. Haüy was unhappy about the merger, fearing that he and his students would be forced to take second place behind the then headmaster of the school for the deaf, Abbot Roch-Ambroise Cucurron Sicard. The merger ended in 1794 after the deaf left the site and Haüy was given the opportunity by the government to re-establish his blind school at the hospice des Catherinettes.[68] But this period was not only marked by disputes with Sicard. Haüy came under severe criticism from a group of young blind people from within his own school. Later that year, they presented a long petition to the National Assembly, complaining about Haüy's cruel behaviour towards them.[69] After this episode, Haüy, who had been appointed the Secretary of the Revolutionary Committee of the Arsenal Section, was labelled a "false patriot" by the Revolutionary Tribunal and even imprisoned for short periods a number of times.[70] In general, scholars have turned to Haüy's ongoing interest in religious ritual and theophilanthropy to explain his trouble with the police.[71] During this period, any form of religious practice was banned and religious leaders were persecuted. Yet Haüy's problems with authorities may have been worsened by the nature of his actual inventions. His communication-objects used to educate his blind pupils were also interpreted by authorities as tools for manipulation.

During the post-revolutionary period, Haüy continued his educational work with the blind without hesitation. His commitment to the notion of "humanity" was now aligned with revolutionary ideals attached to the Declaration of the Rights of Man.[72] When the blind presented their communication-objects, he believed that they were engaging directly with the spirit of liberty, equality and fraternity amongst citizens. Yet Haüy's humane listening ethic now presented a problem to authorities seeking to maintain a control over citizens' behaviour. His communication-objects were now considered *faux bruits*, (discords or false noises), that is, noises consciously constructed by a citizen to subvert the political status quo. The *faux bruit* was an important part of the sensory urban landscape during the Terror. It was used most often in relation to political unrest. In 1792, the Committee of Public Safety reported on the observation made by one of its members that the enemies of the revolution "are spreading the most slanderous rumours amongst the public, claiming that patriots will be arrested and that several of them have already been so; that their aim in sowing these false rumours is to intimidate the good citizens and to rally them around their party to save them".[73] There were many reports during this period of specific incidents of *faux bruits* made by individuals on the street.[74] The phrase also played into broader anxieties about invasive noises caused by revolutionary activity and increasing industrialization. Complaints about unexplained urban noises were very common at this time and were investigated seriously by authorities. Residents were often fearful of new noises caused by new industrial structures (underground tunnels, for example) or artillery (distant cannons, for instance) which they could not see.[75] The emergence of the *faux bruit* attested to a new attitude towards noise that was closely related to post-revolutionary anxieties about violence and terror. "Humane" hearing was replaced by a new, more repressive attitude towards sound, which outlawed those who did not appear to conform to the contemporary political status quo.

Operatic music, as James Johnson has demonstrated, clearly promoted revolutionary and patriotic sentiments in both text and music. Such sounds were sanctioned by members of the elite who attended the opera, paying large sums of money to attend luxurious operatic venues in Paris as part of their public duty. Similarly, the rhetoric of the *Fête Revolutionnaires* as interpreted by Mona Ozouf established "the discourse of the Revolution itself".[76] Haüy's communication-objects, made for the poor blind from the street, were more complex within the culture of revolutionary discourse because they relied on non-verbal communication rather than a worded political message. Though Haüy sometimes relied on texts, he did not believe any verbal message was necessary in his products so long as the sound was presented in a sophisticated way. Sound was the essence of his system. He boasted that his blind students had been promoting the famous "law of equality" in their practice long before it had been formulated on paper by politicians. They did not need empty words to promote "humanity".[77] Haüy pitched sound above language in an attempt to prove his political point. Such an approach was extremely dangerous at this time. Pure sounds were open to charges of *faux bruit*. During the Terror, the appeal of Haüy's system suddenly dissipated. Now, he was charged with corrupting and brainwashing members of the public purely for his own interests, which were opposed to those of the State. Despite his desire to present the blind in more public settings, he was increasingly confined to the cramped quarters of his training sites after the revolution. This made his work appear even more secret and strange. The sound of his blind musicians practising behind the closed doors of a former convent was "overheard" with suspicion by those on the street.

The greatest threat to Haüy's programme, however, came from the blind themselves. Whilst Haüy was complaining to the National Assembly for the disruption to his programme as a result of the merger, some raised the alarm in public about their poor living conditions on the Celestines site. Their petition to the National Assembly began by stating that it was not them, the blind, who needed to be made more civilized through education, but the Legislative Assembly who had wholeheartedly denied them welfare support.[78] The notion that they could make their own living through participation in an educational programme was fanciful, they explained. They then launched into a three page attack on their "leader" who, they explained, "knows well the weaknesses of his means, and saw that such truths, given the light of day, would necessarily eclipse his glory, and overturn his ambitious projects".[79] Haüy had responded in silence to their complaints about their quarters, they continued, and should not be given any extra money for his programme. "Help us to achieve a general law for all blind people,"[80] they added. The young blind ended their petition by asking authorities to take a closer look at Haüy's Institution.

> Enter into this establishment with us; we will be the most confident guides to lead you into this den of quackery. We hope Legislator, that the reasons which earn us the hate of M. Haüy, will be able to earn us the benevolence of the National Assembly. We ask you to agree to our petition so that truth will triumph over the impostor, and so that from now on, the Blind will no longer be the victims of the cupidity of men.[81]

The young blind students who signed the petition clearly believed that Haüy was failing to hear them on an ethical basis. His ego and ambition were clouding his judgment and he was instead exploiting them for his own personal gain, exactly the opposite of his intentions. The petition must have appealed to authorities who now believed in the rights of all, particularly the blind and partially sighted, to a "voice".[82] The National Assembly had recently saved the Hospice des Quinze-Vingts, an institution housing the blind and partially sighted (later to be amalgamated with Haüy's school), from closure on exactly on this basis. "Humanity, justice and the general interest call for the preservation of this previous hospital," a government report concluded.[83] Blind members there had been complaining about treatment by the sighted and calling for the rights of *les malheureux* ("the unfortunate") since 1790. Haüy's young blind students were calling for "humanity" in a similar way. In their opinion Haüy's programme was nothing more than a sign of his ambition and complete absence of care. Their lack of confidence in Haüy was made worse by serious technical problems with his raised musical notation system.[84] Blind students complained that the notes were not embedded deeply enough in the paper and that they simply could not read them. They were being forced to do a pointless task by a "charlatan".[85] The bound books of embossed texts also weighed almost five kilogrammes each. Often in large format, they were difficult to manage. Sometimes, the raised letters on the paper lost their physical shape. Haüy constantly reassured critics that the structured nature of the system was a success.[86] Yet there were clearly serious problems with the new technology and it would be up to Louis Braille to develop a more effective way of making musical scores available to the blind much later, in 1825.

Though Haüy survived the dispute with his students, it placed him in the spotlight with the authorities. His institution was now reported to be a *foyer d'agitation* ("foyer of trouble").[87] There were now suspicions that he was heading a monarchist cult on the site. In 1792, a member of the Revolutionary Committee, Antoine-François Deray, drew attention to "the person named Haüy, former interpreter of tyranny and of the Admiralty, teacher of the Blind, schemer, and false patriot, has sought to make his establishment a foyer of fanaticism, having posted notices throughout Paris that one could attend mass there".[88] Haüy wrote, "I was incarcerated three times as a terrorist".[89] In 1797, Haüy was to hold the first public meeting of the Theophilanthropists at the Institution, complete with mass and moral instruction. Theophilanthropy became a fully authorized sect before it was banned as "Jacobin peril" in 1797. In 1800, the National Institute for Blind Workers was amalgamated with the Hospice des Quinze-Vingts and Haüy was given sole administrative responsibility for the blind students' education there. But the immediate post-revolutionary period had been fraught with problems. It was not simply that Haüy was a Theophilanthropist. His interest in sophisticated sound objects made him the ultimate target of conspiracy theorists. According to authorities, Haüy's work encouraged citizens to attend to a secret voice that was designed to manipulate them. Complaints by the blind themselves that Haüy was failing to care for them only added to suspicions amongst authorities. If Haüy could not even look after his students, how could he be trusted as a democratic thinker? In the eyes and ears of the authorities, he had failed his own test of hearing in an ethical sense.

## The communication-object and economic exchange

After all the problems he had experienced, the challenge for Valentin Haüy was simply to be able to continue his work. In 1800, however, he found himself at the centre of another dispute, this time with the Minister of the Interior, Jean-Antoine Chaptal, who amalgamated

Haüy's school with the Hospice des Quinze-Vingts. Chaptal believed that Haüy's communication-objects could be converted into materials for economic exchange. He saw the new blind school as a kind of work-house for the blind who could support themselves as a segregated group through productive industry.[90] For Chaptal, the communication-object was most "useful" if it was tied to economic productivity. It was not only Haüy who complained about this agenda but the blind them-selves. This time it was the older blind in the institution who did not believe they should work for a pension. In 1802, they wrote to protest against amalgamation with the Institute for Blind Youth. They com-plained that the plan would lead to "perpetual celibacy",[91] since they would have to abandon their wives and children for a life of repression. Haüy complained that the government was inhumane and its intention was to destroy his life's work. In any case, Chaptal implemented a set of rigid police regulations designed to inhibit the blind members' move-ments and behaviour.[92] Haüy's communication-objects were "reduced" to modes of manual work, wool spinning, textile manufacture, tobacco manufacture and print works. Chaptal's intention was to use the blind to produce superior materials for demonstration at the events such as the Industrial Products Exposition and sold throughout the Hospices in Paris. After a long dispute with the Agent General, Haüy resigned, with-drawing to a private school for the blind, the Musée des Aveugles, which he had founded at the time of the Quinze-Vingts amalgamation before leaving for Russia in 1807. He was to take up a new position training the blind in Saint Petersburg on the invitation of Czar Alexander I.[93]

In 1803, Haüy's brother, René-Just, included a complex study of acoustics as part of his *Traité élémentaire de physique médicale* (Elemen-tary Treatise of Medical Physics).[94] It examined sound as the "movement transmitted by percussion, or by any other manner, to the molecules of a body".[95] This was the most highly complex treatise on sound since that produced by Perrault and Nollet and incorporated comments on the mathematical work of Savart and the experiments of Biot. René-Just was interested in how sound was transmitted and its behaviour during transmission. He commented on recent experiments conducted on the speed of sound and echo. He also tried to measure sounds in a concrete way outlining a number of experiments to demonstrate how this might be done, covering topics such as scale and temperament. René-Just Haüy was especially skilled at finding alternative ways to describe sonic effects. He criticized existing models which explained how simultaneous sounds were created without sounding as though they were mixed together, for example. In failing to depict accurately how sounds were actually dispersed, as separate materials which responded to different particles of air instead of rays which intersected, scientists created more confusion, he concluded. René-Just also included a number of radical experiments using strings made of different materials, glass and metal, as well as in different shapes.

René-Just's contribution to the scientific study of sound was closely connected to one of his brother's final piece of work, a new syllabus for teaching reading to the sighted (Figure 4.2).[96] The work made the powerful point that learning to read by learning the alphabet was inefficient, since the letters of the alphabet gave no accurate reference to the sounds of the words: "In our language we have a number of simple sounds for which we have no characters in our alphabet."[97] In his method, Haüy notated all the 43 consonant and vowel-related sounds used in the French language in a table, calling them "sons primitives". He then wrote down all the possible spellings relating to these sounds.

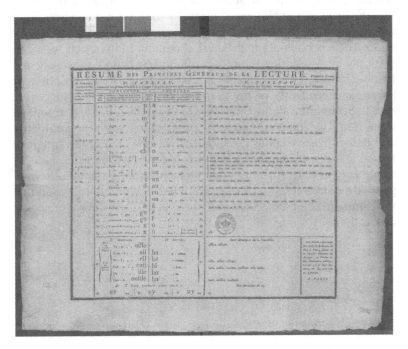

*Figure 4.2* Valentin Haüy, "Résumé des Principes Généraux de la Lecture, Tableau 1 et Tableau 2," *Nouveau syllabaire...manuel de l'élève [et manuel de l'instituteur], par le citoyen Hauy* (Paris: Institut national des aveugles travailleurs, an VIII [1799–1800]). Bibliothèque Nationale de France, Paris

Students would then learn to spell words by memorizing their sound. Haüy's phonetic constructions were, of course, communication-objects in the most obvious sense. They relied on structures of sound to construct objects for study, in this case words and sentences. Sound was used as a tool for demonstration. We now know that by converting words into phonetic sounds, language can be more quickly learnt.

Phonetics is central to the way in which we communicate with each other and express ideas efficiently. But the message of "humanity", embedded within the cultural history of phonetics as communication-objects, has now largely been lost. In blind education, Haüy's work quickly led to the introduction of Barbier de la Serre night writing system and Braille.[98] Yet we have forgotten that as communication-objects, they held a specific meaning relating to their association with the higher listening state. Sound was not merely an ornament in the eighteenth-century space. Its powerful presence in blind education and in reading syllabuses such as the one above made a direct call to the individual to participate in the "humane" ideal. To confront a communication-object was to be transformed into a new socially inclusive state of mind. There is no doubt that Haüy's communication-objects were dramatically altered during the post-revolutionary period, Consulate and Empire. However, Haüy was able to produce a highly effective phonetic system independently of his troubles during this period, suggesting that the communication-object may have survived the test of time in some form. In addition, Chaptal's emphasis on scientific knowledge and design in developing the industrial agenda of post-revolutionary France actually encouraged inventors to construct pieces of auditory technology as part of France's development as an economically competitive nation. Such was the focus on invention and imagination that the communication-object became a beacon of successful technological innovation.

# 5
# Sound, Health and the Auditory Body-Politic

It was not until the Restoration that Haüy's work was reintroduced into blind education. Louis Braille, who entered the institution (now named the Institution Royale des Jeunes Aveugles) in 1819, perfected Haüy's raised-reading system, which had also been developed in earlier years by Charles Barbier de la Serre. Blind students were now encouraged to play an even wider range of musical instruments as a central part of their curricula. The director, Sébastian Guillié reported: "This year they received harp lessons for the first time. It wasn't thought possible until now to teach them this instrument which is so difficult for the clear-sighted themselves, due to the difficult position of the body and the multiplicity of the strings that nothing differentiates."[1] Braille, who became a very successful organist, was to benefit from this revitalized programme.[2] Hearing continued to play a leading role in medical circles throughout the late eighteenth century despite Haüy's departure. The concept of the *faux bruit* had confirmed, rather than weakened, medical suspicions about the need for ethical hearing practices, re-emerging within medical culture as key to restoring a healthy social reality. It is possible to trace, however, a shift in discussions as to how ethical hearing might actually take place in the modern social setting. Audition was no longer relegated to the ear alone, but to the entire human body, which now harnessed its entire inner nervous system. Activating the "auditory body" was now an important way for individuals to become members of the regenerative post-revolutionary body-politic. It connected one citizen holistically to all others within the new nation-state. Haüy had helped forge this shift when he created the concept of the communication-object for his blind pupils. But it was the introduction of new mechanical technologies such as the harp, introduced to the blind as well as young sighted pupils in the early years of the nineteenth century, which consolidated the use of the "auditory body" as a widespread social practice.

During the early nineteenth century, there was a great deal of discussion around the idea that the human body might respond to sound in exactly the same way as the human ear. In 1821, the doctor and surgeon, Jean-Louis Pascal, posed the questions:

> Is this sound a body which really exists in nature, and which would come to strike the ear, or is it simply a way in which the moving body is able to affect us? and in this case, would the sound only be the exercise of the ear applying the knowledge of the movement of the bodies?[3]

The auditory body aimed to construct the social setting in the early nineteenth century through positive, physical action. Because of its auditory nature, music was considered pivotal in encouraging the dynamic auditory body into movement. Whilst certain types of music were important within this discourse (good music produced good bodily responses and vice versa), overall, the application of auditory activity itself was designed to override concerns over musical convention and aesthetic taste. From 1800 to 1830, 80 patents for mechanical inventions built specifically for this purpose were released into the public domain by the Conservatoire des Arts et Métiers in Paris. At first glance, they appear to resemble musical instruments. Yet, on closer analysis, they push the boundaries of musical instrument design to a new extreme. To interpret such patents as communication-objects, with the notion of "humane" hearing at their core, makes logical sense. These were not objects designed to create beautiful melodies or harmonies. They were technological interfaces with sophisticated sound systems, built for the purposes of social reform. They were very different from the optical forms of communication, most famously the Chappe telegraph, that was developed in France and patented a little earlier. It was sound, not vision, that formed the basis of their articulation into space. Today, we can only imagine the sounds that such instruments created. Most, with the notable exception of the *orgue expressif* of Grenié, were never actually built.[4]

Such communication-objects played an important role in refashioning the concept of ethical hearing within modern French society. Some of these inventions eventually became standardized tools for economic consumption and were the source of considerable national pride. Yet it was their sense of sonic sophistication, most obvious at the beginning of the century, that gave them a particular kind of prestige. As we will see, this was closely related to the way in which sound in a sophisticated form was attached to an idealized form of modern citizen. Sonic devices enabled individuals to maintain their sense of social compassion or "humanity" by activating their auditory intelligence and sustaining their bodily health. These qualities, initially at least, were most important to a nation emerging from the instability of revolution. According to some doctors then, sound and its practice remained a critical element of the nation's social and political wellbeing. The concept of ethical hearing had been altered, but it remained present in a highly regarded form of social practice.

## Audition 1780–1830: Le tremblement sonore

In 1778, the French physician, Etienne-Louis Geoffroy, published a lengthy dissertation on the ear.[5] Geoffroy was the son of the famous French doctor and chemist, Etienne-François Geoffroy, whose work on chemical affinity had revolutionized the classification of botanical plants.[6] Geoffroy's work was eagerly awaited by the Société Royale de Médecine and the Académie des Sciences. At the age of 53, Etienne-Louis was yet to acquire the celebrity standing of his father. He had built up a good medical reputation as a physician in Paris. In the natural sciences, he had commenced a global study of snails. It was his dissertation on the ear, however, that gave his career the boost that would make him famous, at least amongst his father's colleagues in the Parisian medical establishment. In 1779, the Société Royale presented Geoffroy's breakthrough findings, most notably that Geoffroy had discovered the binary hierarchy of animal intelligence based on the physiology of the inner ear.[7] At the top of the chain were those living organisms with the full set of auditory mechanisms in the inner ear, the semi-circular canals and the cochlea. These included humans, quadrupeds and whales. In the other lower category were those organisms that had no cochlea but a vestibular apparatus and two or three semi-circular canals. These included birds, reptiles and fish.

In his dissertation, Geoffroy refuted previous theses that reiterated the ancient idea that deafness resulted from faulty eardrums. The test used to demonstrate this theory, based on a model of testing hearing from inside the head, proved nothing, he explained. More conclusive were his observations of the labyrinth, the part of the ear containing the semi-circular canals and the cochlea. These were the most sensitive parts, "intended only to receive it [sound] and to communicate the impression of it to the brain by means of the auditory nerve".[8] Geoffroy concluded that, though the semi-circular canals were important for hearing, the cochlea "must be stronger and more substantial"[9] because it received the force of the entire auditory sensation. He wrote: "The organ of hearing is the only sense which can accurately measure its object. The eyes can indeed distinguish colours but they cannot make out precise degrees between their shades: it is the same for smell and touch. Hearing alone not only distinguishes sounds, but measures them so precisely that one can distinguish the tones, the semi-tones, and their different modifications. This is what gave birth to the art of singing, which is so natural to man that all peoples have had their music, and to harmony which flatters all Hearers even if they are not Musicians. If, like the auditory nerve, the retina had been divided into little cords or fibers of varying lengths, the eye would also be able to measure light, just as the ear measures and distinguishes sounds."[10]

Geoffroy's work on hearing appeared just after the philanthropist and linguist, Haüy, began his work on sonic reform amongst the Parisian blind community.[11] Both posited the importance of hearing, albeit in different ways, as a possible solution to halt human social decline. As Sean Quinlan has pointed out, fears of degeneracy were widespread during the pre-revolutionary period, culminating in the construction of numerous diseases diagnosed by doctors as nervous disorders or diseases of infertility.[12] Quinlan explains: "In many ways, medical practitioners were providing a biological commentary on the perceived social disequilibrium of old regime society."[13] Yet he also demonstrates that such concerns over degeneracy were the beginnings of a much longer movement that extended into the period after the Terror. This movement, which was heightened by the instability of revolutionary events, sought solutions to restoring the larger political economy of the nation-state itself through new "private practices of self-control, rehabilitation and perfectibility".[14] Sound was important in this new movement of health and wellbeing.

Geoffroy and Buisson's work on active listening also aligned with other discussions during this period on the importance of the auditory body. Hearing as a mode of clinical diagnosis, as Laennec had demonstrated, was combined with the idea that the auditory body could be propelled back into health again by activating itself. This was in close keeping with concerns over the instability of the body-politic after revolutionary events. The famous heart surgeon, Marc-Antoine Petit, was to reiterate the pivotal importance of sound on the body when he described its traumatic effects on his patients at the hospital of the Hôtel-Dieu in Lyon during the revolution.[15] Petit, who taught Laennec, described how the noise of cannon fire had permanently altered the bodily states of his patients, explaining that it had led them more rapidly towards death by complicating the nervous aspects of their condition. Petit wove such analysis into a broader discussion on the close link between revolution and bodily ill health. Petit depicted revolutionary activity as a kind of naturally occurring hearing process in society, which would inevitably take place amongst civilized people. The problem with revolution, though, was that bad sounds might ultimately prevail. Long-term social damage, he explained, might be the consequences of short-term revolutionary glory. This was because, during such periods of instability, a nation's sense of bodily equilibrium was compromised, "every passion is in play; every soul is exalted: sensitivity doubles strength; energy is everywhere; and every man feels revulsion at the very idea of an injustice".[16] Petit's understanding of the body-politic as a natural container of carefully balanced sonic relationships was also expressed a year earlier in 1795 in an article in the *Journal de Rouen* entitled *L'orgue politique* ("The Political Organ"). Like Petit, the unknown author believed that revolution had altered, though not permanently destroyed, a nation's humane sound.

> A great nation that we want to regenerate is like an organ that we want to put back together. The artist does not break each pipe which makes disharmonious or discordant sounds: he sets it at the pitch that he desires, and, when he plays the first tune, he delights his listeners.[17]

If citizens could be made to use their bodies to reinstate sonic relationships, a healthy body-politic could be restored.

Bodily listening was not then considered far-fetched as a medical practice. It occurred as an important part of a broader "biomedical program of regeneration" that, as Quinlan suggests moved "from elite clinical theorists to a kind of 'literary underground' of medical practice".[18] The reinstatement of sonic relationships through a kind of self-inflicted form of bodily auscultation was advocated by the Montpellier doctor, Etienne Sainte-Marie, who reinstated the work of the mid eighteenth-century vitalist doctor, also of the Montpellier medical school, Joseph-Louis Roger.[19] In his *Tentamen de vi soni et muscices in corpore humano* published in 1758, Roger accused seventeenth-century natural philosophers of associating the effects of music with magic, explaining that the relationship between music and the body was much more empirically verifiable. This had to do with the sonorous capacity of bodily materials, particularly the nerve fluid in the nerves. In his preface to the

1803 French translation of the work, Etienne Sainte-Marie advocated musical practice as a serious form of bodily hygiene. "Those who look upon music as a purely pleasurable art, will not believe in the beneficial effects on many illnesses that the author of this work attributes to it."[20] He reiterated Roger's belief in the importance of the connection between the auditory and the bodily nervous system, *le tremblement sonore* ("the sonorous trembling") and *le tremblement vital* ("the vital trembling"), explaining that auditory nerves "are spread throughout the body, and dilate it...and awaken in it [feelings of] courage, love, charity, pity, joy, and the expansive passions".[21] And, like Roger, he emphasized that certain illnesses of a nervous nature could be completely cured by rebalancing such a relationship. This included illnesses, Sainte-Marie explained, which were not in essence nervous, but which presented some sort of nervous element such as pulmonary phthisis, as well as deeply embedded emotional illnesses that might result in suicide, and nervous disorders such as hypochondria.

Treatment, Sainte-Marie explained, involved much more than listening to music. It involved action: "A healthy body is not the state which is most conducive to the best effects of music. This art acts more vigorously on a body weakened by some illness."[22] The mechanical effect of sound, the most natural form of exercise containing tiny movements ("which constitute in the organs life itself"[23]) could be fashioned into physical exercise: "Music is an exercise, and must be recommended to women and to people of letters, who lead a sedentary life."[24] The healthiest members of the population were musicians, he wrote: "We note that there are more old people among musicians than amongst other artists."[25] This was because musicians ensured that there was continually vigorous energetic activity in their nervous system. Musicians' bodies were always warmed up since sonic energy was physically retained within the nerves. Sainte-Marie must have advocated Roger's analyses of different sounds and his terms of diagnostic treatment. Roger explained that music was useful: if the patient was a musician or if, in a state of health, he played music a lot; if he was not a musician but demonstrated symptoms of an illness caused by the alienation of the spirit, the alteration of nervous fluid or the excessive tension of nervous fibrils that one recognizes phrenetic delirium; if there was no pain or inflammation; if the patient showed taste and aptitude for music in the middle of the illness; if they showed obvious emotional responsiveness to music; and finally, if there was resistance to normal pharmaceutical remedies.[26] Whilst Roger was interested in specific sounds, the doctor, Forgues, who presented his dissertation at Montpellier a year before Sainte-Marie's translation, explained that there were three main classification groups of sounds which had a physiological effect.[27] These were imitative sounds, sounds that had effect on memory and the particular language of sound of itself. He agreed though, that sound offered man a powerful refuge from disease. It had the capacity to transport the entire self into another domain of existence.

The Montpellier vitalists were not the only medical authorities to advocate the auditory body. Jean-Louis Desessartz, doctor and professor of chemistry and pharmacy at the Faculty of Medicine in Paris from 1776–1778, bemoaned the abandonment of music as a mainstream form of medical practice. Desessartz was a leading expert on health problems relating to climate and had advised the government on the serious health effects created by seasonal changes, as well as specific climactic irregularities[28] He explained: "Why is this art which produces such astonishing effects in healthy men and in sick men, abandoned by doctors today? Why would one no longer have recourse to it for sick people who do not respond to treatment from medications reputed to be the most effective?"[29] Desessartz recounted the case of a woman of 60 who had demonstrated a range of illnesses, catalepsy and convulsion which had not responded to conventional treatment. Yet, after exposure to music, her condition began to improve. Such evidence suggests, Desessartz explained, that the status of certain conventional medical treatments as infallible destroyed the use of successful treatments such as musical listening, which could not be measured on such terms. Doctors were also too impatient to try such techniques, seeking a quick fix solution to the problem, and were also afraid of being ridiculed. Yet according to Desessartz, they often did not understand the proper procedure and care required for musical treatments. Advocates of the auditory body, such as Desessartz, understood sound as an energetic substance that activated the body through the oscillation of fibres. Unhealthy bodies could be gradually coaxed into movement with music, which resulted in cure, as Desessartz explained, *une crise salutaire* (a salutary crisis).[30] This was directly in keeping with miasmatic theories of disease, common amongst hygienists, which blamed contaminated air on bodily irregularities. Since auditory relationships constructed physical environments, sounds could be altered to purify them. As Desessartz, emphasized: "So, everything is matter, and matter in action: this is not an occult quality."[31] Later writers on hearing were to stress more and more the way in which the auditory body engaged in a dynamic way with its surroundings. This led to the replacement of theories on hearing function, such as air implantus used by Perrault and others during the eighteenth century, with much more mechanical explanations of hearing emphasizing the impulsive and interactive auditory nerve.

Audition continued to be portrayed as a tool for personal transformation, even though there was a definite shift towards the human body as auditory receptor and transmitter. Through bodily audition, the individual could engage in preventative and therapeutic health care, which would benefit the body-politic as a whole. This was not necessarily a repressive form of self-expression. Rather, it was designed to draw from the body the most potent source of vitalist power for social purposes. In discourses on bodily audition, the emphasis was on experiment, sensitivity and patience in medical practice rather than surgical or pharmaceutical intervention. Doctors admitted that such diagnostic treatments were difficult to define, yet there was consensus on their medical significance just as at the end of the seventeenth century medical and scientific hearing research were again powerful tools in proving the transformative potential of auditory biology. To use one's ears, and now one's entire auditory body, was not to control the imagination, but rather to momentarily let it loose in an alternative frame of reference. This would shock the self into change and recalibrate the self into a more positive and outward emotional stance towards others. Such was the renewed importance of audition in medical circles that some, such as the military doctor, Pascal, posited in 1821 that more obvious forms of human expression, such as the voice, were, in actual fact, merely reflex actions of hearing.[32] Pascal emphasized the sophistication of hearing and its pivotal effect on the way the body moved in space.

This theme continued in serious medical treatises on otology, such as *De l'oreille* by the doctor, Jules-Charles Teule, published in 1828.[33] Whilst Teule was eager to bring back the fundamental relationship between hearing and the laws of acoustics, he nevertheless stressed the way in which hearing was pivotal to the sense of one's surroundings: "We can also readily accept that the presence of a special hearing organ in a species presupposes a melody, a cry..., as a means of establishing necessary relations between the individual and the species."[34] It was during this period that doctors began to acknowledge the importance of binaural hearing, that is, the way in which two ears work together.

it only requires observation to be convinced of this; taking the velocity of the sound into consideration, one can accept that perception takes place at exactly the same moment for each ear, and that the impressions transmitted to the cause of the feeling by the double organ of hearing, are identical or only differ in their intensity; this does not prevent them from being confused.[35]

Teule also explored the complex issue of how the ear assessed the source of a sound. He explained:

The diversity of the directions of the shaking communicated to the pulp of the auditory nerve. It is a product of education; it is indeed essential that, through multiple comparisons, the mind learns to grasp the connections which exists between the relative positions of the *corps sonore* ["sonorous body"] and the ear, and of the diversity of the sensations of the ear.[36]

Teule also suggested that auditory education was important because sound could so easily disorient an individual by changes in its intensity. For Teule, the auditory body was best practised through music. Music was not presented as an artform but as a method for retaining biologically ingrained sonic values. The "tonic", of the musical scale, Teule explained, was the most important focus for the human *corps sonore*. Yet aspects of rhythm were also present within the body: "It seems that the sentiment of the measurement stems from the actual organization where one sees the movements of the circulation subjected to the power of the rhythm, and consequently, all of the organs [are] aroused periodically and consistently by the blood."[37]

From these observations, Teule concluded that it was the deaf who were the most socially disadvantaged group. They lacked a *corps sonore*. This also explained why they became mute. The blind, however, could be classified as superior in intelligence to the everyday citizen, simply because they could hear at acutely sensitive levels:

People born blind generally make a more reliable judgment than other men on objects for which knowledge has been gained through the ear; it is not rare for a blind person to recognize from the noise of a footstep, people from whom he has been separated for a long time.[38]

Overall, Teule demonstrated that the body consisted of a very real abstract set of sonic relationships, which were best enhanced by music. Music did not simply train the ear but grounded the entire human form in a spatial setting where it could best interact. Doctors reinforced the concept of ethical hearing through a new set of physical parameters. These were overtly centred on the entire human body and grounded in a specific understanding of the biological nature of the *corps sonore*. The auditory body fitted neatly with broader medical codes of social improvement, which stressed individual hygiene and self-improvement. A powerful technical culture appeared in the early nineteenth century as an extension of this medical mode of practice. Machines designed for the auditory body were to be situated in the institution or private home with the very "public" goal of social wellbeing driving their implementation. This shaped the institution and private domain as well as fostering a sense of a unified or "humane" body-politic in a similar way to Haüy's communication-objects for the blind.

## Communication-objects and the auditory body

Patents for the auditory body that were published between 1800 and 1830 might be grouped according to the standard sound propagating materials for musical instruments: string and air.[39] Even this single step towards categorization is misleading, however. The string instruments all elaborate on the traditional harp, piano and violin designs. However, some are such distorted and mixed versions of these models that the traditional prototype is almost entirely unrecognizable. The same dilemma occurs with the air or wind instrument models. Traditional instrument designs as the flute, horn and organ are manipulated to force wind into creating strange acoustical sonorities far from the expected. Yet this group too contains such unusual shapes, not one of which resembles the instruments we know today. In one sense, such objects are the industrial "hardware" products of the acoustical theories of Lagrange, LaPlace and Monge. They demonstrate principles of acoustics by manipulating materials in a variety of different ways. Yet all were constructed with the human body at their centre. They were not mechanical music boxes nor scientific apparatus. Rather, they were bodily machines that simulated modes of ethical hearing.

New designs for harp instruments reflected medical attempts to activate and strengthen the auditory body through a much more accurate and penetrating acoustical sensation. Sébastian Erard's famous chromatic harp with eight pedals was patented during this period.[40] Mechanisms such as a toothed wheel operated by a foot pedal were added so that the body might be more actively employed for the purposes of attaining a sound that is "purer and more intense".[41] The 1802 harp of Michel-Joseph Ruelle relied on a variety of mechanical devices that could be manipulated by the performer in order to achieve such a result. He described his own design for a mechanical peg to keep the string tight, giving an accurate note, and reinforced the harp with iron cladding: "The mechanism is indestructible," he boasted, "when used by experienced hands".[42] There were also mechanisms that allowed the performer to manipulate the speed of sound production: the *pathétique* (prolonging the sound with slow pedal movement) and *agitato* (agitated sound through fast pedal movement). The *harpe d'harmonie* ("harp of harmony") designed by Thory increased the size of the instrument to seven feet and included a number of new sonorous devices: a keyboard producing a combined piano/harp effect, a drum and Chinese bells.[43] Keyser de l'Isle's *harpe-harmonico-forté* added 34 strings strung across the front of the instrument supported by an extra column.[44] Like the Thory harp, the overall aim of the instrument was to enhance resonance using radical mechanical means. Other additions and improvements included the coordination between pedals and levers by Mérimée as well as smaller mechanical improvements by Plane as well as Gilles' long-string lever.[45]

Inventors distorted, reordered and replaced the different components of the piano during this period. The most obvious and famous of these designs was the "Piano-harmonica" of Tobias Schmidt, which added a huge pedal-operated bow above the strings of the piano.[46] Bowing the piano, reviewer's noted, allowed the human ear time to adjust to mixtures of sonorities. By manipulating the instrument in such creative ways, they adapted the small, square and quiet piano to the much more demanding physiological characteristics of the active auditory body. The piano by Charles Côte positioned the keyboard directly over the vibrating table.[47] The human body was therefore fully engaged in a producing a much more harmonious sound. The pianos of Thory, Erard, Cluesman, Pleyel and Dietz were experiments in creating maximum resonance within a compact space.[48] Thory relied on a particular string shape and a forked bridge whilst Erard, Cluesman and Dietz refined the hammer mechanism. Ignace Pleyel's piano included an expanded base for extra resonance, whilst Triquet placed a similar attachment above the vibrating table.[49] The piano mechanism itself was made much more fluid and immediate by Klepfer-Dufaut and Blanchet et Roller.[50] The Blanchet et Roller design relied on the nut of the hammer to push it into action rather than an added pivot. The most famous piano designer of the period, Erard, experimented with new string designs and the shape and form of the vibrating table in dramatic ways. These include the *forté-piano d'une forme et d'un mécanisme particuliers* ("cylindrically-strung piano with the strings moveable by foot") and the *forté-piano ayant la forme d'un sécretaire* ("piano in the shape of a secretaire with a double row of augmented strings also moveable by foot").[51] Eulriot's *piano de forme elliptique* ("elliptical piano") involved the use of an unusual heart-shaped, double-string design. Erard designed a way of regulating the temperature within a vibrating table through a series of metal bars placed between the strings.[52] His design also had a double set of strings operated with the pedal, giving added harmonies. Jean-Baptise Wagner designed a transposing piano with a moveable keyboard stabilized by a "wedge" as well as a sixth pedal that produced added harmonies.[53] Pleyel's *piano unicorde* ("uni-string piano") was designed around the acoustical phenomenon of the single vibrating cord.[54] The focus on the material of the string was also present in Erard's ingenious piano mechanism, *forté-piano à son continu* ("piano with a continuous sound"), which included a sustaining mechanism that secures the hammer to the string via a toothed or rotating wheel.[55]

Overall, harp and piano instruments were transformed into complex acoustical systems emphasizing dynamic aspects of the living body: circulation, vibration and pulsation. The auditory body could be set in motion by either plucking (as in the case of the harp) or hitting the string (as in the case of the piano). Each method achieved equally effective results. Violin makers who concentrated on bowing, of course, experimented with copper and wire strings, string length and mechanisms for maintaining string tension.[56] The resonating case of the violin was tackled in a lengthy patent by Chanot, who provided a marked fingerboard for facility of learning, and dramatic modifications of the "f" hole next to the lower strings of the violin for better resonance.[57] Yet the large harp was most suited to the auditory body because its upright structure most closely reflected the physical and acoustical power of the vibrating human form. The combination of the upright harp and the keyed piano were behind the famous upright piano design of Erard's *forté-piano ayant la forme d'une sécretaire*. Pfeiffer's *harmomelo* anticipated Erard's design simply attached a harp to a set of keys.[58] Pfeiffer explained: "For a long time, we have sought to compose a stringed instrument in a perpendicular position, in order that it can be played with a keyboard; we have always found that the strings tightened in this position produce a stronger, more full, and pleasanter sound than those tightened horizontally."[59] Although the instrument was patented as one imitating the piano, it was strung like a harp inside its case and also included a harp mutation that was added to enhance volume and timbre. The importance of the harp was most obviously displayed in Dietz et Second's *clavi-harpe*, which was again, a kind of keyboard-operated harp.[60] The patent concludes: "The keyboard of this instrument produces, by means of a suitable mechanism, the very same effect as is obtained with the fingers plucking [the strings of] the harp."[61]

The auditory power of the harp was also a central feature in the development of guitar instruments. Mougnet boasted that his *guitare-lyre* emitted sounds that rivalled "those of the harp".[62] Some of these, such as the *guitare-harpe* of Levien were designed to be portable as well as powerful. Others simply increased the number of strings or adapted the guitar design in order to emulate the harp, for example, the *déca-corde*, and guitar design of Lacoux, which ran the grain of the wood alongside that of the string. He explained:

> I obtain, through the analogue placement of the fir veins, a sound which, instead of being over emphasized, as is a criticism of the gui-tar, and resulting in confusion, [the sound] becomes, by vibrating, more round, more harmonious, and closer to the beauty of the sound from the best harps.[63]

The *lyre-organisée* of Led'huy used a relatively small-sized instrument body but with many strings.[64] It had 15 in total, divided into three parts: the bass, the middle range (which also had the option of being muted), and the upper range, which was vibrated through a keyboard mechanism. The inventor explains: "As to *the touch*, it requires much experience and a well-exercised hand to be able to obtain from the keyboard these soft and velvety sounds which please the delicate ear, and which are so capable of inducing emotion."[65] The *guitarion* had a stand and was bowed like a cello.[66] Finally, the enormous *harpo-lyre* of Salomon included a kind of mechanical amplifying box described as a *"corps sonore* or pedestal" operated by pedals (Figure 5.1). The instru-ment was "already much superior" to ordinary guitars "by its size and its shape" and therefore created "at least twice as much sound and a more satisfying harmony". The *corps sonore* addition was added to enhance the auditory body and could be modified by the human performer. Salomon explained that it "encloses a mechanism from which is produced the *crescendo*, the *diminuendo and the tremolo*, plus a bass drum effect using four pedals".[67]

Vital substances of the auditory body could also be activated through the action of breathing. A number of machines focused on manipulating airflow through new instrumental shapes and the addition of keys and slides. The more elaborate brass and wind instruments of today, such as the horn and trumpet of Schmittschneider and the brass instruments of Halary evolved during this period.[68] Others allow the performer extra

*Figure 5.1*   Jean-François Salomon, Original patent manuscript, *Brevet d'invention de 5 ans, instrument de musique, appelé par l'auteur Harpo-Lyre* (Paris, 1829), Cote du dossier: 1BA3120. Archives INPI, Paris

choices in altering air flow. One, such as the *flûte traversière* of Delavena had a detachable key mechanism which left the player with a simple wooden tube.[69] The slide mechanism, which we find today only in the trombone, was applied to the horn and trumpet and even the organ.[70] The *piano éolien* of Keyser substitutes metal strips for keys but is set in motion by air blown into a tube.[71] The most sophisticated of these new instruments was the *orgue expressif* of Joseph Grenié, a model of which was placed in the Paris Conservatoire in 1816. Both women and men had access to the instrument at the institution. Women, whose bodies were considered to contain large amounts hydrogen (and thus required access to normal air)[72] were allowed special permission to enter the male-only division of the building in which the organ was located, to have their special class.[73] This instrument design permitted the volume of the reeds to be manipulated by regulated wind pressure through the use of a double-bellows system.

Musical aides designed to develop the auditory body were also patented. Amongst these were the metronome, which developed from a clock into its triangular box shape.[74] Tuning products included the octagonal metal device that emitted a perfect "A" when struck and placed between the teeth, and a portable mechanical keyboard resembling a fully fledged instrument in itself.[75] Even page turning, which might interrupt the smooth expression of auditory bodily activity, was addressed. The page turner, which was adaptable to all music stands (including keyboard), was a mechanical device operated by foot that enabled reading from the printed score without stopping to turn the page.[76] The musician interacted much more smoothly with the printed score via better printing procedures,[77] the development of reusable manuscript paper which worked like a modern "whiteboard"[78] and a manuscript transposing device.[79] Serre's elaborate educational device, the Chiroplaste, facilitated the musical reading process by indicating the name of the notes over the keys.[80]

Musical culture, as demonstrated by these patents, cannot be understood, therefore, without reference to the auditory body. As serious medical tools, mechanical musical instruments were designed specifically for the human auditory body in action. They were intended to enhance vital fluids, activating the nerves and preventing and healing irregularities in flow that might result in disease. Later in the century, doctors began to warn of the effects of over-using the auditory body. Musicians were considered to be most in danger of hysterical illnesses or nervous disorders as well as other related diseases such as infertility, digestive disorders, anaemia and heart palpitations.[81] The Hungarian ear, nose and throat specialist, Maurice Krishaber, wrote: "No class of society is exposed so frequently to nerve disorders in general, and more particularly to cerebral disorders. Cephalalgia, migraine, lightheadedness, vertigo, insomnia, sensory disturbances, general irritability, hypochondria, and melancholy, are phenomena that we observe in musicians."[82] Krishaber explained that this was not simply due to the powerful effect that music might have on the nervous body. It was also the result of chronic physical overuse in musical practice, warning that many musicians were putting themselves at risk by excessive practice times. Their auditory bodies were cramping up in response, causing blood clots and haemorrhoids. Women organists who practised organ pedalling to excess caused uterine haemorrhages, he explained, making them susceptible to infertility. He also believed that there was a connection between the overuse of wind instruments and pulmonary phthisis. Krishaber consolidated these theories on the nervous body in his book *De la névropathie cérébro-cardiaque,* outlining a neuropathic disease affecting the nerves of sensation and the heart.[83]

The idea of the auditory body also began to shape the urban landscape in important ways. Prior to the revolution, the constructive imagination of those listening was grounded in the individual or group who sought to connect with another realm of experience. Now, it began to be embedded in the physical and technical aspects of the landscape itself. Industrial France resembled an auditory body constantly in a state of renewal through sonic encounter. Such a cultural approach was critical to France's economic advancement throughout the nineteenth century. It enabled inventors to refine acoustical technologies to an extremely high degree and inflect urban spaces with transformative auditory characteristics in creative ways. There were, of course, dangers to such an approach. If the auditory body was to become the social norm then those who would determine sonic law would have unprecedented power. It would be perpetually dynamic, creative, regenerative and productive. This was the reasoning of scientists, musicians and politicians who pledged their interests in the sonorous state.

## Science, industrialization and the auditory body in the nineteenth century

In 1822, the French mathematician and physicist, François Arago, set up a dramatic sonic experiment. Arago fired cannons from the tower of Montlhéry in central Paris and from the Villejuif, on the outskirts of Paris. He used the chronometer, a device recently invented by Bréguet, to measure the speed of sound with much greater accuracy. Positioned at Villejuif, scientists measured the time between the appearance of the visual signal of the cannon's lighting at Montlhéry and their perception of the cannon's sound.[84] Baffled as to why there was not direct reciprocity of results between the two sites, Arago implicated the interference of clouds. The speed of sound was no longer directly proportionate to the source of sound.

Arago's work was, in one sense, linked to Perrault's research. Both imagined a transformative world of sonic objects perceived by the listening ear. Yet, by separating sound from source so obviously, Arago's work emphasized the idea even more forcefully that sound constructed collective space entirely on its own. In his 1825 book, *Géométrie et mécanique des arts et métiers et des beaux arts*, Charles Dupin, the politician and engineer, outlined the advantages of a society shaped by sound. He suggested that it was not human hearing that was most important. Rather, it was the careful utilization of sound and silence that might influence auditory bodies within society to greatest effect. Sound could be used to create efficiency amongst workers in the factory, for example, and soldiers in the military. It should not to be feared by administrators but instead exploited so that its full potential could be harnessed for social good:

> We must not believe that the cadence and harmony of military exercises are a pure object of luxury and parade; they produce the most precious effects. They get the soldier used to adjusting all his movements to the voice of his chiefs, and to the sound of instruments of war; they make one of its organs more compliant using impressions of sound, and consequently more responsive to exaltation and to drill when a successful action is required.[85]

A sonic telegraph for the military was also developed at this time. Tested at midnight in central Paris, the invention consisted of the simple production of musical notes attached to letters of the alphabet. The sounds of the horn projected as the secret code into the night space during military action.[86] The inventor explains: "We have therefore thought that to offer men a new means of communicating their ideas, of transmitting to faraway distances in darkest obscurity, was a real service provided to society."[87]

In 1848 the architect, Théodore Lachez, drew attention to the importance of hearing spaces in the urban environment. The speaking voice was weak, he explained. It required carefully constructed assembly rooms to make it sound.[88] By far the most important demonstration of sound's role in space, however, was in 1850, when the physicist, Léon Foucault began to think about measuring the speed of light with greater accuracy. Previously, the speed of light had only been roughly calculated using the distance between the earth and the sun. Initially, in Foucault's experiment, the central element was Paul Gustav Froment's rotating mirror apparatus, at which a beam of light was directed. It was not until 1862, however, that Foucault was actually able to test his experiment at the Paris Observatory after seeking the help of the acoustician and organ builder, Aristide Cavaillé-Coll. Cavaillé-Coll's bellows consisted of two small alternating bellows that were activated rather primitively by a hand pump. The compressed air was forced into another reservoir topped with a piece of lead, assuring a regulated internal pressure. Cavaillé-Coll also added a regulating box coupled to a smaller second regulator that fed a single wooden pipe. Here was a bellows design that effectively damped down any fluctuations in pressure resulting from the jerky hand pumping motion. The revolving mirror apparatus was attached to the box, so that the mirror revolved extremely fast whilst the pipe sounded. A tuning forked was mechanically connected to the experiment, checking the uniformity of mirror rotation.

Foucault declared that the bellows operated the mirror apparatus at perfect regularity so that the pressure in the regulator did not vary beyond a fifth of a millimetre for every 50 centimetres of air mass. The speed of light was calculated as the beam of light was reflected onto a concave mirror, which returned the light to the rotating mirror. There was a slight displacement between the incident and reflected beams from the spinning mirror. Knowing the speed of rotation and the displacement and path length of the beam, the velocity was calculated at 298,000 km per second, the closest yet to the modern reading of 299,792.5 km per second. Ethical hearing, in this experiment at least, had been replaced by the human hand that pumped the bellows and by the delicate workings of this complex new technology and expertise. It was implicit within the sonic forces of Foucault's machine.[89]

Sonic culture within the nineteenth-century Parisian salon also relied on human-centred messages of transformation which were implicit in their complex technologies. Though salon culture was thoroughly bourgeois, the desire to incorporate socially transformative emotional messages was always present. In the mid-nineteenth century, France was well behind England in its national and international consumption of musical instruments. Published statistics reveal that between 1852 and 1857 the ratio of instrument consumption in England to that in France was 100 to 11. In a similar statistical result, the ratio of instrument importation in England to that of France was 100 to seven.[90] What these figures do not show, however, is that the French were also developing and promoting mechanical organ instruments during this period at a high rate.[91] These instruments, which can be loosely described as organs, were based on Grenié's basic model of *orgue expressif* from 1816, mentioned earlier.[92] Major product names of these instruments included the orgue expressif (of various different models), the mélodium, the orgue-Alexandre and the harmonicorde. Statistics show that patents for this type of instrument increased dramatically between 1851 and 1860[93] and, as in the piano industry, there was also a marked increase in numbers of patrons; a 30 per cent increase between 1847 and 1860.[94] Their development in the mid-century period helped to prevent a serious national disgrace when French piano firms Pleyel and Erard, and the emerging Bord piano firm were commercially overtaken by successful

German and American piano firms like Steinway and Bechstein at the end of the nineteenth century.

Because of their complexity, these instruments were initially very expensive to produce and for the individual to buy. Even famous composers and performers such as Franz Liszt and Frédéric Chopin, who adored these instruments, were only able to help stimulate the occasional purchase by a very wealthy buyer.[95] Advertisements for such instruments often omitted the cost of the instrument, merely directing the potential customer to the warehouse for viewing.[96] The lack of actual trade, however, did not necessarily lead to lack of interest. Whilst the Erard and Pleyel piano companies were successfully exporting and setting up companies abroad, there were vast numbers of high-profile national and international concerts involving these organ instruments. Such concerts were not just forms of arbitrary entertainment. They were blatant advertisements for new sonic forms of hardware custom-built for the domestic environment.

In 1856, the French music journal, *La Revue et Gazette Musicale de Paris* described in detail one small musical performance by an unknown musician, Judith Lion, at the salle Herz in Paris. On 28 February, Lion had premiered a new organ "reed instrument", the *harmonicorde*, developed by the Debain firm, which "combined the single string and resonating reed producing extremely novel piano/harp effects". This instrument was based on the same design principle as Grenié's *orgue expressif*, but was slightly more compact in its construction. The anonymous reviewer of the concert recognized not only the quality of the performance or the performer, but the quality of the instrument and its potential for human aural stimulation, writing that "[its extraordinary power] distinguishes this instrument from the rest through its variety of timbres that cannot be found in any instrument of the same type. The imitation of the diverse instruments of the orchestra is pushed to a point where it can no longer be surpassed."[97]

Contemporary reviewers of these technical "sound instruments" stress the desirability of the instrument through a variety of terminologies. Adolphe de Pontécoulant described the Debain harmonium through its ability to express transformative emotion. It is, he wrote, a

> reunion of many diverse instruments, and reveals and directs the sensibility; that the affections of the soul are endowed with new life by the various effects of many instruments carefully combined: a reunion of instruments is very difficult to find, and we have searched for an instrument to replace them all.[98]

But Pontécoulant also referred to the actual technical feature of the prolonged acoustical experience that could only be achieved by its complex mechanical design. He wrote: "The piano lacks the length of sounds. The pianist, deprived of this magic and mysterious power that penetrates the heart so deeply, desperate because of his lack of power, combats fiercely in finding other ways of exciting the public."[99] Cutting-edge acoustical inventions of this kind were the most effective piece of public communication, Pontécoulant believed, because they connected with the people.

Cyril Ehrlich has explained that the decline of the international French piano trade was implicitly linked with "international disgust" at the French piano exhibits at the Paris Exhibition of 1889 and 1900.[100] These expensive, over-designed, wooden and iron framed vertically strung exhibits were, according to international onlookers, a poor match for the Germany's medium-class instruments, which used low-cost iron frames in a grand piano format. Yet, in France, the organ industry was in the midst of a peak period in relation to the international marketplace. Joseph Merklin, who had developed many forms of *orgue expressif* during the mid-nineteenth century, developed and displayed the first electrically powered organ, completed a year before his triumphant display of two electro-pneumatic organs at the 1889 and 1900 *Expositions Universelles*. Merklin's Gold Medal, followed by his Grand Prize, was won "fair and square" with the blessing of a number of American admirers. The French journal *Le Monde Musical* described the scene: "Electrical keyboards, enshrouding the thousands of people, were the object of great curiosity, each person could experience the effect produced by these marvelous process of transmission and they applauded this new victory of science."[101] The new era of electrically powered French organ design and trade emerged in the context of a nation suffering from a dramatic decline in its international piano trade. With the mass shipping of electronic Merklin organs throughout Europe, America and further afield to Guadeloupe, Martinique and Monaco in the 1890s, French organ design became one of the nation's most high-profile commercial products, at a time when France was increasingly worried about its commercial image on the international stage.[102] Hence, the auditory imagination continued to shape critical aspects of French modern history.

# Conclusion

This book demonstrates that within eighteenth- and nineteenth-century French culture hearing did not simply relate to what was "heard" in a literal sense. Rather, it occurred in relation to cultural debates and discussions surrounding the need for individuals to conceptualize new universalizing social structures for the modern age. During this time, France maintained the same transformative model of the extension of the self that was used in the pre-Enlightenment Catholic past. However, the vocabulary surrounding such a model was situated squarely in the secular domain of social reform, rather than the theological realm of divine morality. Because of the way in which hearing knitted the private authorial listener and the public collective "voice", it became a powerful resource for the imagining of new material sites that might attend to people's hearing needs and so offer a model of "humanity". The Châtelet provided the public with one of the most important material sites of "the humanity of hearing" within the urban space. It was ultimately tested according to the way in which it attended to the everyday citizen's complaints and used as a model for the social reimagining of a more democratic judicial space.

Hearing also offered medical practitioners a means of reinterpreting the hospital space and the clinic as a site of protective care and humane consideration. Bordeu emerged as one of a number of French medical physicians, including Buisson and Laennec, who reimagined the body through its resonant qualities alone. Pre- and post-revolutionary culture was marked by the use of sound to reimagine bodies and to situate those bodies within specific institutional sites coded by the sounds of new technologies. The blind body was used to demonstrate such sonic reimagining. Blind musical expression was part of a broader sonic ideology, extending well into the nineteenth century that positioned the individual citizen-body into a collective body-politic.

The way in which hearing forces us to reconfigure a whole series of historiographical questions relating to the history of modern France is a central conclusion of this study. Indeed, what this study demonstrates is that modern French society as it evolved throughout the eighteenth and nineteenth centuries relied on hearing models from the past precisely because they offered what vision could not: a way to refashion social spaces. Society at that particular time required radical conceptual models to construct itself just like the "space-time" resonant world offered by the model of the divine Universe. This could be mapped onto the secular social world using a moral and ethical vocabulary relating to "humanity" and "human rights".

The idea of fragility within the figure of the modern French self is opened up for critical examination when considering cultures of hearing. Duverney's resonant ear demonstrates fragility in so far as the world in which it presents is materially fragile and subject to disease and decay. Yet even within the more "objective" scientific and medical discourses, hearing in terms of its function and meaning remains a mechanism that deliberately confounds. As Duverney explained, "I admit that it is difficult to understand how this works: they are mechanical movements which are imperceptible, and it is very difficult to explain their nature and causes."[1] Duverney and Perrault were well aware that sound perception does not occur in a linear temporal way. Rather, like the sonorous relationships within the divine universe, sound processing within the human ear constructs space by necessarily turning time in on itself. The way in which hearing might demonstrate a more holistic form of self, a kind of "diapason subject",[2] as Nancy puts it, operating within eighteenth- and nineteenth-century culture and society, suggests that certain modern forms of self, at least in France, were not as fragile as we might think. In this study fragility and fragmentation (as during the revolution) occur only once the ethical hearing model breaks down. Hearing offers a very different model of the self to the sensationalists, for example, who used sensory training to counteract the imaginative potential of the fragmented self. The individual who "heard" had the potential to provide solutions to a fragmented society by engaging fully with their imagination.

The reimagining of urban space in late eighteenth-century France through attention to hearing models is also important. This is achieved within scholarly work not so much through the analysis of specific sound sites (though these are also important) but through close examination of the rich culture of urban restructuring surrounding social reform and renewal. The closed model of the panopticon must be redrawn with alternative, more open models incorporating emotional affect and communication into interpretative analysis. Such open models more keenly appropriate hearing as a foundational element of construction. However "expert" and "rational" they might have appeared, modern medical spaces were imbued with a form of emotional investment related to a culturally constructed consideration towards the sound of the patient's suffering and the patient's "need". Literal interpretations of the gaze have sometimes obscured this important feature of medical and scientific culture in eighteenth- and nineteenth-century France.

Hearing offers us useful material to reconstruct an alternative narrative to the simplistic "public replaces private" argument surrounding the transition from divine rule to democratic government in modern France. Because hearing combines both an authorial hearer and the incorporation of a public form of "voice", it acts as a means through which we can interpret the waxing and waning of private versus public need as political remodelling took place. Sonic technologies in late eighteenth- and early nineteenth-century France were central to the new ways in which the private individual related to the "public" body-politic. The maintenance of a culture of the auditory imagination through cultures of invention after the revolution has significant implications in terms of understanding the construction of the "imagined" French nation which expanded through invisible networks across the globe throughout the nineteenth and twentieth centuries.[3]

How can we assess how such a particular culture of hearing has shaped our modern-day world generally? There is no doubt that its containment within the technical and scientific apparatus of industrialization has ultimately caused a kind of crisis of its own. That is, that it is much more difficult than ever before to locate the listening experience within the networks (now global in reach) of social and political connectivity as Malebranche, Fénelon, Duverney and Perrault conceived of it. How do we ascertain if the "humanity of hearing" has really taken place? In contemporary discussions on health, however, listening as an essential "humane" task is making a slow and steady comeback. Health professionals of all different kinds are now acknowledging that the ethical hearing model must be placed at the forefront of discussions on wellbeing. They explain that hearing at many levels of control provides the key to a healthy global body-politic. The resurgence of interest in the idea recalls the central relationship between healthy individuals, institutions and societies that our French history of ethical hearing has emphasized. In many ways, contemporary research echoes the sentiments of politicians, lawyers and doctors in eighteenth- and early nineteenth-century France who fought for a revitalized listening relationship as a solution to various forms of social need and wellbeing. Like their predecessors, the most powerful of these proposals are grounded in a call for human responsiveness to be more consciously built in to pre-existing structures of care and compassion. Good, active listening, involving the construction of sound itself, remain central features of our modern landscape.

Such social and political constructions of ethical hearing are today powerfully ratified by science and medicine, just as they were in the past. Auditory science, the term now used to describe the discipline, actually covers a whole range of quite distinct scientific fields that have evolved from historical work examining how we hear.[4] These oscillate between those that focus purely on the ear itself (otology, Duverney) and those examining the auditory scene (psychoacoustics, auditory scene analysis, Perrault). There is no doubt, however, that the field of auditory neuroscience has given us a new dimension to human hearing. Yet work on the auditory brain also recalls pre-existing meanings of eighteenth-century scientific and medical work. Sets of results produced on the role of the human auditory cortex, for example, reflect in one way the historical practice of fashioning technical representations of the auditory imagination in the mould of Duverney's striking ear designs. The auditory neuroscience of today falls back on the transformative meaning of hearing as identified by Enlightenment philosophers. It is the "conversion of the mechanical energy carried by airborne sounds into bioelectrical signals, and coding the information content of those sounds by frequency filtering".[5] The enormously complex sets of electrophysiological data used to explore this principle stand in the same historical tradition of technical sophistication as a means of engaging the constructive imagination. Recent textbooks re-emphasize the "fascinating complexity of the auditory brain" which receives "cues that are extracted from the spectral, temporal and level differences at the ears that arise from sounds occupying different locations in space".[6] This definition of hearing perpetuates the idea of hearing as a superior means of social and political survival, just as it did in the Enlightenment.

Themes that are integral to the concept of the auditory imagination continue to be widely pursued in auditory science circles. There is a fascinating pool of recent work on active listening, which emerged, as we have seen, as a central concept in the work of the early nineteenth-century doctor, Mathieu-François-Régis Buisson. Scientists have recently demonstrated that the neural auditory system completes this process, not simply by attending to signals, but by actively ignoring them.[7] Listeners actively ignoring a sound "free up" their auditory cortex through a frontal lobe/auditory cortex rapid auditory displacement process. The auditory brain selects and rejects certain signals in order to function most efficiently. The desire for the most effective rhetoric as demonstrated by such eighteenth-century promoters of the auditory imagination as Bernard Lamy, continues to be revisited, albeit with a striking degree of technical expertise. Auditory scientists work on ways to "clean up" their modern physiological recording systems by blocking out "glitches" produced by environmental noises (power lines, machines etc.), physiological noises (heart, muscle activity etc.) and sensory noise (transducer or electronic noise). Sensor noise suppression, for example, is achieved through a double projection process, effectively cancelling out defects. As researchers explain:

> The value of recorded data in scientific or clinical applications depends critically on the level of noise. Noise narrows the range of conclusions that can be drawn from experimental data, and makes them less reliable. New applications such as brain-machine interfaces (by which a handicapped person can control a machine) are still limited by noise and artifacts, and significant progress in noise-reduction techniques might lead to a breakthrough in those applications. Every effort to reduce noise is worthwhile.[8]

Like the Enlightenment philosopher Fénelon, we have recently turned to the animal kingdom to locate a complex yet self-sufficient listening world as a model for our own. Projects such as the Elephant Listening Project at Cornell recognize the sophisticated role of sound in the creation of local social structures that have survived over a long period of time. Though we cannot hear the low-frequency noises that elephants make to communicate with one another over long distances we now know that they take place. Revealing sound models such as these helps us to imagine a topographical society for ourselves with humanity at the core. Statements used to capture the project emphasize the importance of animal models in relation to our own.

> They are more like us than we think...Elephants are long-lived, intelligent creatures with excellent memories and complex social relationships...What we couldn't hear was always there. Elephants use a variety of calls to express emotion, keep in touch and pass-on

information. ELP discovered that some calls are too low for humans to hear.[9]

The so-called voice of nature is constantly providing considerable material for scientific analysis. Recent recordings of glaciers are only one example.[10] Sound art has also allowed us to express the fragility of the listening experience through the projection of artificial voices in a new and striking way. Digital audio and video signals (which are also "noise-tolerant") have given us the ability to construct the concept of ethical hearing, just as Valentin Haüy did when he created his phonetic and touch-based communication-objects with the new materials of industrial France.

The importance of listening within the human political landscape is constantly revisited. It provides clues to the success or failure of political structures that we have created. There has been recent emphasis, for example, on the exact nature of the listening relationship between state and nation in the US: "Federal systems provide ample opportunities for their policy systems to interact. Indeed, perhaps Morton Grodzin's (1966) most significant contribution to the study of federalism was his observation that federal systems behave as a structure with many cracks. When policy is impeded or opposed at one level, it readily flows to another."[11] Yet, as such studies demonstrate, ascertaining the level of "attention" that one system has with another often or directly with the voting public, often produces unsatisfactory results. National policy systems, as David Lowery and others demonstrate, do not readily interact with those of the state despite their structural interdependence. Voters in Europe, James Adams shows, respond to changes in party images but not to their policy statements: "Overall, our findings thereby suggest that Left-Right ideology *does matter* to voters and that they react to parties' perceived ideological shifts. But, because voter perceptions do not track the parties' actual policy statements, there is a disconnect between shifts in elite policy discourse and voter reactions."[12] Attempts to identify alternative listening patterns independent of the political status quo are increasingly emphasized as significant. Radio-listening clubs, for example, are seen as an important means of democratic expression for youths in Malawi even if they do not incorporate direct political discussion.[13]

Studies on listening also inform scholarship aimed at improving institutional frameworks outside politics. These extend to those institutions, such as the Institutional Criminal Court, which extend beyond national boundaries. There is now an important recognition of hearing directly from a victim within the criminal courtroom environment.

Hearing from the victim involves a different communication dynamic from learning about the impact of the crime through the sentencing submissions of the prosecutor. The adversarial system creates an antagonistic dynamic between the accused or offender and the prosecutor that may well undermine the effectiveness of communication ... the victim sensitizes the offender to the effects of his or her conduct on other people.[14]

Julian Roberts studies exactly how a victim felt about the presentation of the victim impact statement after it has been "heard". He also acknowledges: "Judges in particular, report finding victim impact statements to be a more useful way of learning about the seriousness of the crime."[15] Yet there is also caution regarding the power that this auditory form of expression might have over the formal process of sentencing and parole as a whole. The victim's ability to create a *faux bruit* has influenced scholarly debate surrounding the ways victim's voices should be "heard".

Indeed, the way in which "proper" listening might be integrated into our increasingly complex social environments is a constant feature of consideration. Schoolrooms are constructed as listening environments in order to integrate different types of ethnic, racial and class-based belief systems. Teachers are encouraged to listen to students as a way of understanding students' interests and home contexts. This is then fed back into classroom tasks which might otherwise focus only on standard or generic forms of curricula. Teachers are also encouraged to "learn to listen for the rhythm and balance of a classroom"[16] so as to improve on teaching routines. Listening is used in this context in much the same way as it was during the Enlightenment. It is a way of thinking about how a classroom might be expanded or opened up to accommodate all the needs of those using it. As educationalists explain:

The urban-focused teacher education program from which the study participants graduated attempts to prepare teachers to teach within the curriculum mandates of urban classroom *and* to take on the role of activists committed to transforming classroom practice through a deep understanding of children, pedagogy and curricula. The goal is to prepare teachers to face the challenges of *what is* and also to imagine *what might be*.[17]

Such meanings are ideally suited to discourses on human health. Hospitals and citizen-patients were originally some of the most significant and powerful listeners because they relied on the message of hope that was intrinsic to listening's design. Interest in sound continues to inform research on clinical structures, auditory health, hospital environments and our ageing state of mind. There is a resurgence of recognition that sound is what draws us together as healthy beings and that sound must be attended to, protected and preserved at all costs. Most importantly, sound is now beginning to enter discussions on the management and governance of health across a range of local, national, transnational and global environments. Researchers suggest that structures and projects which fail to incorporate the proper hearing processes will never be successful in the way that patients and those in need really require. Thus, the healthy listening space as outlined by Tenon continues to inform our understanding of the way our compassionate and modern society ought to operate.

A medical health team in Holland recently established, for example, an Active Listening Observation Scale (ALOS-global). The team wanted to prove that there was a quantifiable link between active listening within a standard doctor/patient encounter and patient wellbeing. The study acknowledged from the outset that listening had been seriously overlooked in recent health discussions even though "patients themselves seem to value the personal, active and listening doctor most".[18] The numerical scale was constructed through analysis of the results of a questionnaire given to patients about their GP's listening skills. For researchers, such skills involved body language and facial expressions, since these were powerful indicators of interaction as well as language consisting of the verbal acknowledgement of the patient's feelings. Researchers wanted to demonstrate that active listening involved a whole variety of different skills

> bearing instrumental and affective significance. On the one hand, active listening is an important instrument for gaining information, e.g. by the use of open ended questions, summaries and clarification. On the other hand, it signifies the acknowledgement of a patient's suffering. What is more, the very act of listening assumes that there is something to listen to, i.e. that the patient has the opportunity to talk and express himself. Active listening, therefore, incorporates verbal as well as non-verbal facilitation of patient talk.[19]

The gestural as well as verbal nature of active listening as defined by these contemporary researchers recalls the identification of the listening individual in eighteenth-century circles as a dynamic vehicle for expression. Listening is not simply achieved by sitting silently and attentively near a speaker. It often involves the construction of complex verbal and bodily signs that resonate amongst those around them. One of the most interesting aspects of the study was the use of a patient anxiety index in conjunction with ALOS-global. Patients were asked before and after the consultation how they were feeling, in addition to questions from the listening scale. Researchers found that there was a high correlation between the anxiety of the patient and the level of active listening during initial visits. More anxious patients also demanded more non-verbal modes of active listening. Their reliance on the proper-functioning auditory body as early nineteenth-century doctors understood it, as much as the listening ear, became much more urgent depending on their state of mind.

ALOS-global demonstrates that listening still represents one of the most profoundly transformative aspects of the human condition, and that it is still considered key to the creation of a compassionate society. Embedded in the listening scale are the unstable and fragile characteristics that late seventeenth-century doctors and scientists also highlighted in their tracts as essential to the hearing process. This instability, in the case of ALOS, generated as much by anxious, suffering patients as the affective verbal and non-verbal signs of the listening process itself, provides the cathartic experience that forges a resolution to the social divide. There is an essential blurring of boundaries between the individual and the collective that listening as a creative process of the constructive imagination can instil. Behind the scientific language is a realization that medical environments which depend on uneven levels of power and control will only work successfully if they highlight what is essentially a ritualized and deeply historicized act of human confrontation and renewal.

Reconstructing sonority in order to recalibrate the mind and body is also becoming an accepted aspect of mainstream health discussion. Tinnitus, the modern medical term for any noise in the ear, is now treated with a complex form of sound therapy. The condition, which Duverney firmly acknowledged in his *Traité de l'ouïe*, is also associated with hyperacusis, a negative reaction to sound depending on its physical characteristics, and the relatively new terms, misophonia, a negative reaction to a sound with a specific pattern or meaning. Rather than mask sound directly, doctors have found a way to manage unwanted noise by habituating the ear to a low-level broadband alternative.[20] This enables the negative associations with tinnitus perception to be short-circuited.

Auditory retraining was, of course, implicit in Valentin Haüy's late eighteenth-century curriculum for his blind students. The idea that the perceptive ear could be modified through focused exposure to a purer form of sonority was implicit in his work. He extended sonic therapy to touch-based technologies (such as embossed text), which were perceived by the body as a sound system drawing the blind individual into a collective group. All Haüy's technical materials can be described as communication-objects, material things containing systems of sounds. Political concern over the *faux bruit* during the revolution caused doctors to draw on sonority again for solutions to its perceived ill effects. The concept of the auditory body, which could actively propel itself into health through musical activities, was a widely accepted solution. The auditory body in motion was, in actual fact, an extension of Haüy's communication-object, though it embraced much more openly the entire physiological body including its internal nervous systems.

Faulty signals from external sources are also under scrutiny today in the same way. Scholars are beginning to acknowledge the effects on cardiac patients of unwanted noise from various technologies within ICU units.

> To maintain and monitor their vitals, they [patients] are attached to many different instruments and devices, such as a respirator, electrocardiograph, blood pressure meter, central venous pressure meter, pulmonary artery catheter, and so on. Noise generated by these instruments and devices becomes a potential stressor and brings negative effects to patients.[21]

In order to counter such effects, some researchers have offered patients musical intervention treatment whilst they are in critical care. Results demonstrate that during the first postoperative day, there is a decreased rate of "noise annoyance, heartrate, and systolic blood pressure regardless of the subject's noise sensitivity".[22]

The importance of hearing amongst our ageing population is also a focus of attention. Researchers found that "seniors with hearing loss are significantly more likely to develop dementia over time than those who retain their hearing".[23] Investigators also suggest that there might be a "common pathology" between hearing loss and dementia. The relationship between hearing and an intelligent mind is one that nineteenth-century doctors would also have validated, though obviously through very different means. More broadly, listening is now a central feature in our discussions on the governance of health. This includes work on a whole range of health issues from the mobilization of non-governmental organizations (NGOs) in the global health arena to the introduction within a democratic nation of new cutting-edge medical technologies.[24] The way in which hearing might expand our understanding of knowledge on health and science and technology by facilitating the incorporation of voices falling beyond the limits of traditional expertise is a constant theme. The humanity of hearing remains critical to our modern way of life.

# Notes

## Introduction

1. François Fénelon, *Les aventures de Télémaque [par Fénelon]* (Paris: Vve. de C. Barbin, 1699). All translations are from François Fénelon, *Telemachus*, ed. Patrick Riley (Cambridge: Cambridge University Press, 1994).
2. Fénelon, *Telemachus*, 7.
3. Ibid.,108.
4. Ibid., 332.
5. Veit Erlmann, *Reason and Resonance: A History of Modern Aurality* (New York: Zone, 2010).
6. Erlmann quotes Foucault, *The Order of Things: An Archaeology of the Human Sciences* (New York: Vintage Books, 1994), 63.
7. Erlmann, 68.
8. Jan Goldstein, *The Post-Revolutionary Self: Politics and Psyche in France, 1750–1850* (Cambridge, Massachusetts: Harvard University Press, 2005), 16.
9. Erlmann, *Reason and Resonance*, chap. 2.
10. Sophia Rosenfeld, *A Revolution in Language: The Problem of Signs in Late Eighteenth-Century France* (California: Stanford, 2001).
11. Sophia Rosenfeld, "On Being Heard: A Case for Paying Attention to the Historical Ear," *The American Historical Review* 116, no. 2 (April 2011), 328.
12. Lawrence Klein, "Enlightenment as Conversation." In *What's Left of Enlightenment? A Postmodern Question*, eds. Keith Michael Baker and Peter Hanns Reill (Stanford: Stanford University Press, 2001), 150.
13. Jürgen Habermas, *On the Pragmatics of Communication*, ed. Maeve Cooke (Cambridge: Polity, 2003).
14. Dena Goodman, *The Republic of Letters: A Cultural History of the French Enlightenment* (Cornell: Cornell University Press, 1996), 97.
15. Dena Goodman, "Public Sphere and Private Life: Toward a Synthesis of Current Historiographical Approaches to the Old Regime," *History and Theory* 31, no. 1 (February 1992), 14.
16. James Johnson, *Listening in Paris: A Cultural History* (Los Angeles: University of California Press, 1996).
17. Alain Corbin, *Village Bells: Sound and Meaning in the French Countryside*, trans. Martin Thom (New York: Columbia University Press, 1998).
18. Joseph-Guichard Duverney, *Traité de l'organe de l'ouïe* (Paris: Estienne Michallet, 1683).
19. Erlmann, *Reason and Resonance*, 94.
20. Michael Bull and Les Back, eds., *The Auditory Culture Reader* (Oxford: Berg, 2003); Veit Erlmann, ed., *Hearing Cultures: Essays on Sound, Listening and Modernity* (Oxford: Berg, 2004); Mark Smith, ed., *Hearing History: A Reader* (Georgia: University of Georgia Press, 2002); Jonathan Sterne, ed., *The Sound Studies Reader* (London: Routledge, 2012).

21. Sterne, *Sound Studies*, 2.
22. Walter J. Ong, *The Presence of the Word: Some Prolegomena for Cultural and Religious History* (New York: Global Publications, 2000). First published in 1967 by Yale University. See Ong's many other publications which touch on oral culture.
23. Ong, *Presence of the Word*, 117.
24. Ibid., 146.
25. Ibid., 163.
26. Penelope Gouk, *Music, Science and Natural Magic in Seventeenth-Century England* (Yale: New Haven, 1999), 18.
27. Ibid., 266.
28. Jonathan Sterne, *The Audible Past: Cultural Origins of Sound Reproduction* (Durham, NC: Duke University Press, 2003), 2.
29. For details of this tradition see Fenner Douglass, *The Language of the French Classical Organ: A Musical Tradition before 1800*, new and expanded edition (New Haven: Yale University Press, 1995). See also my discussion of sound and nineteenth-century French culture in *Women, Science and Sound in Nineteenth-Century France* (Frankfurt: Peter Lang, 2007).
30. Jean-Luc Nancy, *Listening*, trans. Charlotte Mandell (New York: Fordham University Press, 2007).
31. Ibid., 12.
32. Ibid., 13.
33. Ibid., 17.
34. Ibid., 22.
35. Jacques Attali, *Noise: The Political Economy of Music*, trans. Brian Massumi (Minneapolis: University of Minnesota Press, 2009), 6.
36. Georgina Born, *Rationalizing Culture: IRCAM, Boulez and the Institutionalization of the Avant-Garde* (Los Angeles: University of California Press, 1995).
37. Daniel Pressnitzer, "Auditory Scene Analysis: The Sweet Music of Ambiguity," *Frontiers in Human Neuroscience* 5, no. 158 (December 2011), 1–11. doi:10.3389/fnhum.2011.00158.
38. Other well known examples include Roland Barthes, Pierre Bourdieu and Gilles Deleuze.
39. Here I am referring in particular to Foucault's writings from 1954 to 1969. See Michel Foucault, *Dits et Ecrits*, trans. Daniel Defert and François Ewald, vol. 1 (Paris: Gallimard, 1954). Translations are taken from Michel Foucault, *Death and the Labyrinth: The World of Raymond Roussel*, trans. Charles Ruas (London: Continuum, 1986) and Michel Foucault, "Introduction," trans. Donald F. Bouchard and Sherry Simon, in Gustave Flaubert, *The Temptation of Saint Anthony*, trans. Lafcadio Hearn (New York: Modern Library, Random House, 2001), xxiii–xliv.
40. Foucault, "Introduction," xxvi.
41. Claude Perrault, *Essais de physique; ou, Recueil de plusieurs traitez touchant les choses naturelles*, vol. 2 (Paris: Jean Baptiste Coignard, 1680).
42. Randall McGowen, "Power and Humanity, or Foucault among the Historians." In *Reassessing Foucault: Power, Medicine and the Body*, ed. Colin Jones and Roy Porter (London: Routledge: 1994), 107.

43. "le médecin écoute et interprète." Michel Foucault, "Message ou Bruit?" In *Michel Foucault: Dits et Ecrits 1954–88*, vol. 1, ed. Daniel Defert and François Ewald (Paris: Gallimard, 1994), 557.

# 1   Medicine, Science and the Auditory Imagination

1. Blaise Pascal, *Pensées*, trans. A. J. Krailsheimer (London: Penguin, 1995), 4. Krailsheimer's translations are primarily based on the text, Pascal, *Œuvres completes*, L'Intégrale (Paris: Seuil, 1963).
2. For sources on the development of the theory of sympathetic resonance see Penelope Gouk, *Music, Science and Natural Magic in Seventeenth-Century England* (New Haven: Yale University Press, 1999) Alain de Cheveigné, "Pitch Perception Models – A Historical Review," Paper presented at International Conference on Acoustics, Kyoto, 2004; R. Bruce Lindsay, "The Story of Acoustics," *The Journal of the Acoustical Society of America* 39 (1966), 630; Békésy, Georg v. and Walter A. Rosenblith, "The Early History of Hearing – Observation and Theories," *The Journal of the Acoustical Society of America* 20, no. 6 (November 1948), 736–7.
3. See Gouk, *Music, Science and Natural Magic*.
4. See Erlmann's important discussion of Cartesian listening in *Reason and Resonance*.
5. Nicolas Malebranche, *The Search after Truth*, trans. and ed. Thomas M. Lennon and Paul J. Olscamp (Cambridge: Cambridge University Press, 1997), 11. Lennon and Olscamp's translations are of Nicolas Malebranche, *De la recherche de la verité*, trans. Geneviève Rodin-Lewis, in Malebranche, *Œuvres complètes de Malebranche* (Paris, J. Vrin, 1958–1984).
6. Ibid., xxxvi.
7. Ibid., xxxvi
8. Ibid., 301.
9. Ibid., 10.
10. Ibid., xlii.
11. Ibid., xlii.
12. Ibid., xli.
13. Pascal, *Pensées*, 13–14.
14. Ibid., 187.
15. Ibid., 35.
16. Ibid., 60.
17. François Fénelon, *Traité de l'Existence et des attributs de Dieu*, in *Œuvres philosophiques de Fénelon*, nouvelle édition collationnée sur les meilleurs textes (Paris: Charpentier, 1843).
18. François Fénelon, *Réfutation du système du Pére Malebranche sur la Nature et la Grace*, in *Œuvres philosophiques de Fénelon*, nouvelle édition collationnée sur les meilleurs textes (Paris: Charpentier, 1843).
19. "Le jour est le temps de la société et du travail: la nuit, enveloppant de ses ombres la terre, finit tour à tour toutes les fatigues et adoucit toutes les peines; elle suspend, elle calme tout; elle répand le silence et le sommeil; en délassant les corps elle renouvelle les esprits. Bientôt le jour revient pour rappeler l'homme au travail, et pour ranimer toute la nature." Fénelon, *Traité de l'Existence et des attributs de Dieu*, 14.

20. "une voix souveraine et toute puissante." Ibid., 45.
21. "la sagesse supérieure." Ibid., 23.
22. "C'est un maître intérieur qui me fait taire, qui me fait parler, qui me fait croire, qui me fait douter, qui me fait avouer mes erreurs ou confirmer mes jugements: en l'écoutant, je m'instruis; en m'écoutant moi-même, je m'égare... Le maître qui nous enseigne sans cesse nous fait penser tous de la même façon. Dès que nous nous hâtons de juger, sans écouter sa voix avec défiance de nous-mêmes, nous pensons et nous disons des songes pleins d'extravagance." Ibid., 54.
23. "Pendant qu'il me corrige en France, il corrige d'autres hommes en Chine, au Japon, au Mexique et au Pérou, par les mêmes principes." Ibid., 55.
24. Ibid., 40.
25. Sylvain Matton, *Trois médecins philosophes du XVIIe siècle* (Paris: Honoré Champion, 2004), 103–38.
26. "soit respecté des autres Corps comme le Roi d'Univers." Guillaume Lamy, *Discours anatomiques, Avec des Réflexions sur les Objections qu'on lui a faites contre sa manière de raisonner de la Nature de l'Homme, et de l'usage des parties qui le composent* (Rouen: Lucas, 1675), 2.
27. "éclat orgueilleux" Ibid., 5.
28. "un ouvrier qui a tout fait pour soi-même, n'ayant point de fin plus noble à se proposer. Il a produit la matière avec des mouvements sans ses différentes particules, par la nécessité desquels tous les corps que nous voyons et une infinité d'autres qui nous sont inconnus, ont été formés." Ibid., 29.
29. "J'appelle Bruit l'effet d'une agitation particulière que la rencontre dans l'air voisin, et presque en même temps dans un plus éloigné, et jusques dans l'organe de l'ouïe." Perrault, *Essai de physique*, vol. 2, 5.
30. Erlmann, *Reason and Resonance*, 79–80.
31. "la matière dans laquelle se fait l'impression de la forme du son. Cette matière consiste en deux genres de parties, dont unes sont les nerfs dilatés et mêlés avec une substance propre et particulière à l'organe de chaque sens. Les autres sont celles qui sont absolument nécessaires à la fonction de l'organe immédiat." Perrault, *Essai de physique*, vol. 2, 221.
32. Erlmann, *Reason and Resonance*, 80.
33. "On ne dit pas proprement le son d'un canon, d'une caresse, ni d'un moulin, parce que ces bruits ne sont point de l'espèce désignée par le mot de son, qui signifie une espèce de bruit dont la durée surpasse celle du coup qui l'a produit." Perrault, *Essai de physique*, vol. 2, 1.
34. "libéralité" and a "magnificence d'esprit." Ibid., 2.
35. "L'agitation qui fait le bruit le plus souvent ne touche que l'oreille, et ne cause aux autres corps les plus mobiles, aucune émotion sensible, quoi qu'elle fasse impression sur l'oreille à une très longue distance..." Ibid., 6.
36. Ibid., 5–10.
37. Ibid., 10–17.
38. Ibid., 17–35.
39. Ibid., 35–66.
40. "une voix qui fait un long cri." Ibid., 68.
41. "l'oreille en est frappée avec une force tout-à-fait extraordinaire." Ibid., 74, 72–8.
42. Ibid., 78–100.

43. "Il pourrait se faire que les vibrations des corps sonores placés dans l'eau se communiquant, comme l'expérience le fait connaitre, par les parties de l'eau même, fissent leur impression sur quelque partie destinée particulièrement à les ressentir et à les distinguer, et cette partie pourrait avoir toute autre conformation que celle de l'oreille, et être placée partout ailleurs qu'à l'endroit où l'on trouve cet organe dans les animaux terrestres. De cette manière les Poissons peuvent avoir un sentiment très-vif du bruit et des sons qui se passent dans le milieu qui leur est propre, et dont il leur importe le plus de connaitre les différentes modifications." 24 avril 1743, "Mémoire sur l'ouie des poissons et sur la transmission des sons dans l'eau," *Histoire de l'Académie royale des sciences avec les mémoires de mathémathiques et de physiques tirés des registres de cette Académie* (1746), 199–224, 223–4.

44. Abbé Nollet, *Leçons de physique expérimentale*, vol. 3 (Paris: Guerin, 1745), 395–501.

45. "Les vibrations d'un corps sonore se passeraient dans un parfait silence" he explains, "s'il n'avait entre lui et nous quelque matière capable de recevoir et de transmettre cette espèce de mouvement: car tel est l'ordre de la nature, qu'un corps n'agit point sur un autre, s'il ne le touche par lui-même ou par quelque matière interposée, et de tous ceux qui ont imaginé des exceptions à cette loi générale, on peut dire qu'aucun n'en a encore donné des preuves suffisantes." Ibid., 409–10.

46. Guillaume Lamy, *Explication mécanique et physique des fonctions de l'âme sensitive* (Paris: Lambert Roulland, 1678).

47. "Il n'est pas vrai comme on pense dans la philosophie ordinaire, que ce que nous ressentons, soit dans l'objet qui excite le sentiment. La chaleur que le feu produit chez nous n'est point en lui-même, non plus que la douleur n'est point en l'aiguille qui nous pique; mais comme l'aiguille est tellement figure qu'elle peut quand sa pointe nous excite ce que nous appelons piqûre ou douleur." Ibid., 11.

48. See Perrault, *Essais de physique*, vol. 1, (Paris: Coignard, 1680), iii and Lamy's "Addition curieuse." In *Explication mécanique* (Paris: Roulland, 1681) and (Paris: Laurent d'Houry, 1683 and 1687).

49. Lamy was also attacked by the Church for his beliefs.

50. "En effet, comment savoir les intérêts des Princes? comment se gouverner dans les matières de Religion? Et comment juger les procès si l'on ne sait l'anatomie de l'oreille, puisqu'il faut avoir des oreilles pour tout cela?" See Préface, Lamy, *Explication mécanique* (Paris: L'Houry, 1687).

51. "puisque dans le commun ni dans celui d'aucune science, on ne saurait dire que le son de la voix, d'un luth, d'une cloche, soient un bruit. Au contraire, quoi qu'il soit vrai que dans le langage vulgaire, on ne dit pas le son du tonnerre, d'un carrosse, etc. il est certain qu'en langage Philosophique, si l'on demande à quel genre se rapporte le bruit d'un carrosse, d'une porte, etc. on répondra fort à propos que c'est à un genre de qualité que l'on appelle son, à peu près de même manière qu'en langue vulgaire, on ne dira pas qu'un homme est un animal, si on ne veut lui faire injure." See "Addition Curieuse," Ibid., 389.

52. A. C., "Du Verney." In *Dictionnaire encyclopédie des sciences médicales*, première série, A–E. vol. 30, eds. Amédée Dechambre et Léon Lereboullet (Paris: Asselin, 1864–1888), 729–31.

53. Cheveigné, "Pitch Perception Models."
54. Ibid.
55. Békésy, "The Early History of Hearing," 731–34.
56. Penelope Gouk, "Music and the Nervous System in Eighteenth-Century British Medical Thought" (unpublished chapter, 7 February 2012). 14.
57. Ibid.
58. Cheveigné, "Pitch Perception Models."
59. "La petitesse et la délicatesse des parties, qui le composent, renfermées, comme elles sont, dans d'autres parties, dont la dureté est presque impénétrable, rend leur recherche pleine de beaucoup de difficultés, et leur structure a quelque chose de si embarrassé, qu'il n'y a pas moins de peine à l'expliquer, qu'il y a à la découvrir." Duverney, *Traité de l'ouïe*, ii.
60. "ce n'est pas assez que des figures soient vraies et fidèles, si elles ne sont encore faites et disposées d'une manière, qui en ôte toute l'ambiguïté, j'ai représenté les parties de l'Oreille droite, toujours en leur situation naturelle, et pour conserver les premières idées qu'elles donnent, et ne les point laisser embrouiller et détruire par d'autres." Ibid., iii.
61. "peau mince et délicate, garnie particulièrement dans les jeunes sujets, de quelque peu de graisse"; "une enveloppe nerveuse." Duverney, *Traité de l'ouïe*, 2.
62. Ibid., 17.
63. "une espèce de châssis auquel en dessous est appliquée et collée une membrane de même que le papier huile est appliqué sur le châssis." Ibid., 24–5.
64. "Le calibre de ces canaux est quelque fois rond et quelque fois ovale et il s's'élargit vers leurs extrémités comme le pavillon d'une trompette," Ibid., 35.
65. "fait deux tours et demi autour d'un noyau." Ibid., 31.
66. Ibid., 40.
67. "plusieurs filets qui se distribuent à tous les pas de la lame spirale." Ibid., 46.
68. Ibid., 78–9.
69. "parce que les vibrations de ces petits cercles sont plus promptes et plus fréquentes." Ibid., 103.
70. "Je n'attacherai point aux divisions que les Auteurs en font ordinairement, mais je suivrai ici, comme j'ai fait dans l'explication des usages, l'ordre de ma description." Ibid., 110.
71. "une sensation fâcheuse." Ibid., 117.
72. "Après cela doit-on s'étonner si les douleurs du conduit sont si cruelles et si violentes." Ibid., 123.

## 2 The Juge-Auditeur and Hearing the People

1. For the formative text on the invention of "human rights" in the context of the Revolution see Lynn Hunt, *Inventing Human Rights: A History* (New York: Norton, 1997).
2. See Foucault on individual liberty in *The Final Foucault*, ed. James Bernauer and David Rasmussen (Cambridge: MIT Press, 1987), 19.
3. See in particular Jeremy Caradonna's excellent work on the transition from virtue to rights, "The Monarchy of Virtue: The Prix de vertu and the

Economy of Emulation in France, 1777–1791," *Eighteenth-Century Studies* 41, no. 4 (Summer 2008), 443–58.

4. "Mémoire pour justifier le droit du Sceau du Châtelet, attributive de Juris-diction par tout le Royaume, 10 Xbre 1437." AD/II/7. *Archives Nationales de France* [hereafter AN].

5. Hoffbauer, *Paris: A Travers les Ages, Aspects Successifs des Monuments et Quartiers Historiques de Paris: Depuis le XIIe siècle à nos jours, deuxième édition* (Paris: Firmin-Didot, n.d.), 93–5.

6. See for example the study by Richard Mowery Andrews, *Law, Magistrancy and Crime in Old Régime, Paris* (1735–1789) (Cambridge: Cambridge University Press, 1994). This does not discuss the jurisdiction of the Juge-Auditeur. The most comprehensive description of the Juge-Auditeur's courtroom is in Charles Desmaze, *Le Châtelet de Paris* (Paris: Didier, 1870).

7. "c'est par l'exemple le maitre, le domestique, le manœvre ou le pau-vre locataire et propriétaire de la maison, l'étranger et l'aubergiste." "Mémoire sur la proposition de réunir l'office du juge-auditeur à celui des Commissaires-Examinateurs: Et les avantages que le public recevra." n.d., [after 1683?], M.21574., fol.84–85, n.d. Bibliothèque Nationale de France: Archives et Manuscripts [hereafter BNF AM].

8. "légères et d'un modique intérêt." Michèle Bimbenet-Privat, *Série Y: Châtelet de Paris* (Paris: Centre Historique des Archives Nationales, n.d.), 12–13.

9. See for instance, the listing of the Jurisdiction des Auditeurs under the head-ing *gens du Roi, Almanach Royal* (Paris: D'Houry, 1725). See Achille Luchaire, *Manuel des Institutions français, période des capetiens directs* (Paris: Hachette, 1892), 575–6, for discussions of *gens du roi*.

10. "Les auditeurs avaient donc trois caractères distinct: ils étaient juges, enquê-teurs et administrateurs." Luchaire, 576.

11. See *Lettres patentes du roi Philippe le Bel portant defences aux auditeurs* (Saint-Denis, 18 December 1311).

12. "grand détriment de nos pauvres sujets." *Déclaration du Roi qui ordonne les Auditeurs du Châtelet de Paris jouiront les Droits…* (Paris, 1572), 11–12.

13. "Les Auditeurs du Châtelet de Paris jouiront les Droits, Privilèges, auxquels ils sont accoutumés, et dont jouissent les autres Conseillers dudit Châtelet." Ibid.

14. "jusqu'à la somme de vingt-cinq livres." Ibid.

15. *Déclaration du Roi…Augmentation de pouvoir et droits aux Auditeurs desdits Châtelet* (Paris, 1683).

16. Bimbenet-Privat, *Série Y: Châtelet de Paris*, 12–13.

17. See Chambre de l'Auditeur, Minutes de sentences, 1691–1791, Y 8248–406. AN.

18. "OUÏ le dit M. Chavignan lut son plaidoyer." *Minutes de sentences* (2 January 1770), Y 8363. AN.

19. "Edit pour les auditeurs du Châtelet," (2 December 1553), M. 21574, fol. 73; "Déclaration en forme d'Edict pour les Auditeurs du Châtelet de Paris," (6 July 1572), M. 21574, fol. 74; "Qui ordonne que les Auditeurs du Châtelet de Paris," (16 July 1572, 9 April 1576), M. 21574, fol. 75; "Conseilleur du Roi," M. 21574, fol. 76, n.d. BNF AM.

20. For a sense of this world, see Arlette Farge, *Vivre dans la rue à Paris au XVIIIw siècle* (Paris: Gallimard, 1992), 21.

21. "Le Service du Roi et le bien public demandent que les commissariats soient remplis de sujets capables d'attirer la confiance et la considération du peuple." "Mémoire sur la Proposition de réunir l'office du Juge Auditeur à celui des Commissaires examinateur Et les avantages que le public recevra." n.d., [after 1683?], M. 21574, fol. 84. BNF AM.

22. "secours toujours prompt et présent pour pacifier ses petits différents de tous les jours." Ibid.

23. "Juge ou pacificateur, ces deux termes laissent au commissaire la possibilité de choisir entre un grand nombre d'attitudes dont il est en définitive le seul maître. C'est là que les sources surprennent et étonnent, bouleversant peu ou prou les schémas bien construits. Le commissaire, image classique de la répression, prend soudain de nouveau traits." Farge, *Vivre dans la rue*, 220.

24. "Picard." *Sentences et procès-verbaux divers* (January 1791), Y 8407. AN.

25. "1785, M. Picard, Juge Auditeur, rue Sainte-Croix de la Bretonnerie." *Almanach Royal* (Paris: D'Houry et Debure, 1791), 393.

26. "Louis-Claude Picard: Juge-Auditeur au Châtelet de Paris, 16 mars 1785." *Lettres de provision d'offices*, G–J, 1785 V/1/520. AN.

27. Herni Gerbaud and Isabelle Foucher, "Série Y: Châtelet de Paris et prévôté d'Île de France," (Paris: Archives Nationales de France, n.d.), 2. See also the "Le procureur de la commune a requis la demolition du Châtelet, 9 Sep. 1792," *Reimpression de l'ancien moniteur*, vol. 13 (Paris: Plon Frères, 1847), 641.

28. "Mémoire présenté à l'Assemblée Nationale par le Ministre de l'Intérieur, 17 oct 1791." In *Publications sur la Révolution de Paris*, vol. 7, ed. Alexandre Tuetey (Paris, 1905), 101–2.

29. See for example, "Pétition d'une partie des huissiers à Paris à l'assemblée nationale à l'effet de protester contre le décret de 29 sep. 1791." Tuetey, *Publications*, vol. 7, 102.

30. "Question à juger sur la liquidation de l'Office du ci-devant Juge-Auditeur de Paris," *Procès-Verbal de l'Assemblée Nationale, liquidation et remboursement*, vol. 2, AD/XVIIIc/69, n.d., 1. AN.

31. "Il n'obtiendra que le prix que son office lui a couté; il n'obtiendra que de quoi rembourser ses créanciers à qui il appartient tout entier. Ce qu'il doit aux amis qui l'ont aidé, ce qu'il doit aux sentiments d'honneur dont il surtout donne des preuves par son attachement à la révolution, lui commande cette démarche auprès de la restauration de la Patrie." "Question à juger sur la liquidation de l'Office du ci-devant Juge-Auditeur de Paris," Ibid., 3.

32. "Ayant appris le mauvais état où est à présent le Châtelet de Paris ... et étant d'ailleurs touché des misères que souffrent ceux qui sont détenus ... tant pour le peu d'espace ... que pour l'humidité et obscurité des logements, infection et mauvais air des cachots; ce qui cause beaucoup de maladies fâcheuses à ceux qui les habitent quelque temps." Hoffbauer, *Paris: A Travers les Ages*, 101.

33. "Plan du Châtelet de Paris, avec des changements et Augmentations, levé en 1676, Bruand, Libéral (architecte)" N/III/Seine/476/1-3, AN. See also Hoffbauer, *Paris: A Travers les Ages*, 100.

34. Robert de Cotte, Agence Jules Hardouin-Mansart, *Plan du Châtelet de Paris* (dessin) (Paris: Ca. 1685), Bibliothèque Nationale de France: Estampes et Photographie [hereafter BNF EP].

35. Hoffbauer, *Paris: A Travers les Ages*, 101.

36. "Cette démolition projetée sous l'ancien régime, procura l'avantage de déboucher la rue de Saint-Denis, de faire jouir les citoyens qui l'habitent de la vue du Pont-au-Change, et de rendre plus pur l'air infecté par les vapeurs de la Morgue, du Marché aux Poissons et des boucheries." "Le procureur de la commune a requis la demolition du Châtelet, 9 Sep. 1792," 641.

37. "Parmi une foule de traits intéressants, ou au milieu de la colère et de la vengeance du peuple, on aime à retrouver sa bonté naturelle et le sentiment des principes éternels de la morale et de l'humanité." Ibid.

38. "Innocent ou coupable, leur dit-il, je crois qu'il serait indigne du peuple de tremper ses mains dans le sang de ce vieillard." Ibid.

39. "Nous avons vu avec plaisir démolir ces tours du despotisme. Nous avons examiné avec horreur les cachots énormes pouvant renfermer un bataillon de victimes; heureusement que tout est comblé...la rue de Saint-Denis a un débouché avantageux; en outre les habitants ne respirent plus un mauvais air." Hoffbauer, *Paris: A Travers les Ages*, 100.

40. "les vielles murailles moussues du Châtelet recélaient des êtres humains à l'agonie, ayant en perspective un affreux supplice. De simples accusés y étaient soumis à la torture par des juges inflexibles comme Rhadamanthe, y gémissaient, y croupissaient au fond des vrais sépulcres, d'où s'exhalaient, des cris douloureux." Hoffbauer, *Paris: A Travers les Ages*, 93.

41. "Si vous avez dans une maison un endroit sale, obscure, fétide, malpropre, plein d'ordures, les souris et les rats s'y logent infailliblement. Ainsi dans la fange et le chaos abominable de notre jurisprudence on a vu naitre la race rongeante des procureurs et des huissiers." Mercier, Tableau de Paris – Chapitre CCVI, éd. Jean-Claude Bonnet, "Procureurs, Huissiers," vol. 1 (Paris: Mercure de France, 1994), 513.

42. "Notre jurisprudence n'est qu'un amas d'énigmes prises au hasard dans les ouvrages de quelques jurisconsultes d'une nation étrangère; et quand les coutumes et les lois différentes sont privées de clarté, ne vous étonnez pas des monstruosités de la procédure." Ibid., 514

43. "grassement dans le labyrinthe de la procédure." Ibid., 513.

44. "Il y a huit cents procureurs, tant au Châtelet au Parlement, sans compter cinq cents huissiers exploitants, et tout cela vit de l'encre répandue à grand flots sur le papier timbre." Ibid., 515.

45. "ces agents subalternes." Ibid., 518.

46. "Le Parlement est la principale source de tous les abus qui se commettaient." *Lettre contenant quelque reflexions sur les abus de l'administration de la Justice à Paris & un premier moyen d'y rémedier, enrichie de bonnes notes* (London and Paris: Monmoro Librairie, 1789), 2.

47. "Parce que l'ordre étant établi et rendu public, les Procureurs ne peuvent point ignorer quand l'affaire dont ils sont chargé doit être plaidée." Ibid., 19.

48. "Parce qu'enfin la Justice ne sera plus un labyrinthe inconnu et inconcevable." Ibid., 20.

49. "La Société doit sûreté et tranquillité à tous, et justice à chacun; il faut donc que tous les Citoyens puissent aisément se plaindre, que l'on puisse s'assurer d'un homme sur des soupçons, mais que l'on ne juge que sur une conviction complète. Police exacte, sans inquisition, justice humaine et publique, peines douces mais inévitable, voilà le système des pays libres." Adrien Duport, *Principes fondamentaux de la Police et de la Justice, préséntés au nom du Comité de Constitution* (Paris: Baudouin, Assemblée Nationale, s.d., 1789–1791), 8–9.

50. Ch. L. Chassin, *Les Elections et les Cahiers de Paris en 1789: Documents Recueillis mis en ordre et annotés*, vol. 2 (Paris: Jouraust et Sigaux, 1885), 321.

51. See "Question à juger sur la liquidation de l'Office du ci-devant Juge-Auditeur de Paris," 4.

52. Chassin, *Les Elections et les Cahiers*, vol. 2, 315, 321, 328, 382–85.

53. "Séance du soir, M. Picard a fait lecture d'un memoire 18 jul. 1789," *Reimpression de l'ancien moniteur*, vol. 1 (Paris: Plon Frères, 1850), 604.

54. "les caresses les plus insidieuses et les mensonges les plus hardis, pour les empêcher de suivre le mouvement de leur Cœur qui les porterait à remplir les devoirs de citoyens pour la défense de la patrie." Ibid.

55. "Proces-Verbal de la Séance du mercredi, 22 jul. 1789," *Reimpression de l'ancien moniteur*, vol. 1 (Paris: Plon Frères, 1850), 611. See also Chassin, *Les Elections et les Cahiers de Paris en 1789*, vol. 3, 504, 561, 649–50.

56. "Juges de paix," *Table Alphabétique des Lois Insérées dans le Bulletin des Lois de la République française, 1794–95*, 22. See also Marc Deffaux, *Commentaire sur les Justices de Paix* (Paris: Librairie de Jurisprudence de Cotillon, 1838).

57. *Lettre de M. Picard, Avocat et Juge Auditeur; a MM. les membres de l'Assemblée Nationale, Formant le Comité de Constitution.*

58. "mais croyez qu'il n'est pas moins dangereux de donner ce droit de souveraineté (car c'est vraiment un droit de souveraineté que celui de juger en dernier ressort) à des juges de première instance, quelque modique que soit la demande, et quelque précaution que l'on prenne, lorsque la demande sera plus importante, soit pour augmenter le nombre, soit pour mettre plus de soin et de discernement dans le choix des juges." Ibid., 2.

59. "La prévention, l'inattention si naturelles à l'homme, je ne parle pas des passions qui corrompent l'âme, je ne parle que des nuages qui offusquent l'esprit; tout veut, tout crie à l'humanité faible et fragile qu'il est possible qu'un premier jugement soit une erreur, s'il n'est pas une injustice." Ibid., 3.

60. This is explained in Dufey, *Des Auditeurs ou Essai Historique et Critique sur les Revolutions de l'Ordre Judiciaire* (Paris: Brissot, 1828).

61. Ibid., 16–18.

62. "Les conseillers-auditeurs travaillent a rendre la justice; ils ne formaient dans les cours qu'une faible minorité; ils soulageaient les autres magistrats des fonctions plus pénibles; leur part dans l'avancement était réglée, et on avait toujours eu soin de la diminuer." *Observation sur le projet de loi relative à la suppression des Conseilleurs-Auditeurs, Chambre des Pairs*, 22.

63. Deffaux, 1838.

64. See François Furet's discussion in the introduction to *The Old Regime and the Revolution*, by Alexis de Tocqueville, ed. François Furet and Françoise Mélonio, trans. Alan S. Kahan (Chicago: University of Chicago Press, 1998).

## 3 Hearing and Spaces of Medical Care

1. Michel Foucault, *The Birth of the Clinic* (Oxon: Routledge 2003), xvi.
2. Thomas Osborne, "On Anti-Medicine and Clinical Reason." In *Reassessing Foucault: Power, Medicine and the Body*, ed. Colin Jones and Roy Porter (London: Routledge, 1994), 34.
3. Ibid., 35.
4. Foucault, *The Birth of the Clinic*, 141.
5. Ibid., 33.
6. Ibid., 142.
7. Michel Foucault, "Message ou Bruit?" In *Michel Foucault: Dits et Ecrits 1954–1988*, vol. 1, ed. Daniel Defert and François Ewald (Paris: Gallimard, 1994), 557–60.
8. "le médecin écoute-interprète." Foucault, "Message ou Bruit?", 557.
9. See the excellent chapter on Laennec by Jonathan Sterne, "Techniques of Listening." In *The Audible Past: The Cultural Origins of Sound Reproduction* (Durham, NC: Duke University Press, 2003) chap. 2.
10. Abbé de Recaldé, *Traité sur les abus qui subsistent dans les hôpitaux du Royaume, et les moyens propres à les réformer* (Paris: Barrois, 1786).
11. "Sous un Règne qui comble la Nation de bonheur et de gloire, on doit être étonné d'entendre de toutes parts les plaintes et les gémissements des malheureux que renferment les Hôpitaux" adding that, "le plus précieux pour l'humanité est sans doute celui qui procure à l'homme malade et pauvre, un asile dans lequel il peut guérir." Ibid., 1–2.
12. "le plus beau de tous nos sentiments: ce sentiment intime qui nous attache à notre semblable." Ibid., 2–3.
13. "de la frivolité, de luxe, de débaches." Ibid., vi.
14. "l'intérêt personnel." Ibid., 7.
15. "cri de l'infortune." Ibid., 4.
16. "la classe la plus malheureuse." Ibid., 30.
17. "On a toujours reconnu que la classe la plus malheureuse, était cependant celle la plus précieuse à l'Etat pour la culture, les arts et la population: cette idée n'est pas nouvelle…On ne doit donc pas laisser dépérir la partie robuste qui, parmi nous, supporte le plus rude fardeau, qui rend le plus de services à la Société, et à laquelle nous devons tant de reconnaissance et de commisération." Ibid., 30.
18. "Les tables des Officiers dans tous les Hôpitaux sont très bien servies, ils y ont des logements commodes, et dans plusieurs les Pauvres n'ont que du gros pain, peu ou, point du tout de viandes, et de mauvais légumes." Ibid., 25.
19. "On voit en effet que les Trésoriers, Receveurs, Economes, mènent une vie délicieuse au sein de la misère; ils font voir tous les jours aux malheureux, aux dépens desquels ils vivent, la terrible différence de leur abondance avec leur détresse." Ibid., 23–4.
20. "Ma faible voix, dénuée d'éloquence, vous appelle au secours des malheureux." Ibid., 111.
21. Ibid., 2.
22. Dora B. Weiner, *The Citizen-Patient in Revolutionary and Imperial Paris* (Baltimore and London: John Hopkins University Press, 1993).

23. For a more modern example of this see Alice Street, "Affective Infrastructure: Hospital Landscapes of Hope and Failure," *Space and Culture* 15, no. 1 (2012), 44–56.

24. For an over view of the "Problem" of the Hôtel-Dieu see Phyllis Allen Richmond, "The Hôtel-Dieu of Paris on the Eve of the Revolution," *Journal of the History of Medicine and Allied Science* 16 (1961), 335–53.

25. The version I refer to is Jacques Tenon, *Mémoires sur les hôpitaux de Paris* (Paris: Imprimerie de Ph.-D Pierres et Royez, 1788).

26. "c'est le sanctuaire de l'humanité." Ibid., i–ii.

27. "on y est reçu à toute heure, sans acception d'âge, de sexe, de pays, de religion; les fiévreux, les blessés, les contagieux, les non-contagieux, les fous susceptibles de traitement, les femmes et les filles enceintes y sont admis: il est donc l'Hôpital de l'homme nécessiteux et malade, nous ne disons pas seulement de Paris, et de la France, mais du reste de l'Univers." Ibid., i.

28. "il n'est plus de proportion entre la Ville, ses environs et leur infirmerie, le pauvre y est pressé, quatre et six couchent dans le même lit." Ibid., iii.

29. "Quelle méthode suivre dans ces recherches pour se conduire utilement?...Il s'agissait de l'homme, et de l'homme malade: sa stature règle la longueur du lit, la largeur des salles; son pas, moins étendu, moins libre que celui de l'homme sain, donne la hauteur des marches, comme la longueur du brancard, sur lequel on le transporte, détermine la largeur des escaliers d'Hôpitaux." Ibid., ix–x.

30. These are outlined in Husson, *Etude sue les hôpitaux* and Eugène Boisseau's article, "Hôpitaux, Hospices." In *Dictionnaire encyclopédique des sciences médicales*, ed. Amédée Dechambre, série 4, vol. 14, HER–HYG (Paris: G. Masson and P. Asselin, 1888), 275–425.

31. *Mémoire sur la nécessité de transférer et reconstruire l'Hôtel-Dieu de Paris, suivi d'un projet de translation de cet hôpital proposé par le sieur Poyet* (Paris, 1785).

32. "La disposition des salles terminées aux deux galeries circulaires, qui font une communication générale, nous a paru assez bien entendue. Cette disposition est infiniment préférable à celle de l'Hôtel-Dieu actuel, où les salles sont accouplées, à celle même de la plupart des hôpitaux, dont les salles s'enfilent réciproquement, et où l'air en circulant peut porter dans l'une ce qui sort de l'autre." *Extrait des Registres de l'Académie Royale des Sciences du 22 novembre 1786, rapport des commissaires chargés, par l'Académie, de l'examen du projet d'un nouvel Hôtel-Dieu* (Paris: Imprimerie Royale, 1786), 93.

33. "Il faut que les citoyens trouvent au moins dans ces rassemblements un abri contre les intempéries de l'air; il faut qu'au moins aucun malaise physique n'y vienne troubler leurs plaisirs et l'explosion de leurs sentiments patriotiques." Bernard Poyet, *Projet de cirque national et de fête annuelles* (Paris: Migneret, 1792), 10. See also Poyet, *Projets de places édifices à ériger pour la gloire et l'utilité de la République* (Paris, 1799–1800).

34. "Nous louerons l'auteur du Mémoire, qui a si bien plaidé la cause de l'humanité, et nous dirons encore que le projet même de cet hôpital mérite des éloges, que sa disposition est bien entendue et remplit son objet à beaucoup d'égards, que cette construction aura une grande supériorité pour la salubrité, pour la commodité des malades et la facilité du service sur l'Hôtel Dieu." *Extrait des Registres de l'Académie Royale des Sciences du* (22 novembre 1786), 96.

35. "Jamais sujet ne fut plus digne de fixer l'attention d'une Compagnie savante; tout le rend recommandable: son objet, le vœu du Souverain, l'empressement du public, le mérite qu'il y aurait à surmonter les difficultés sans nombre qu'il présente." Tenon, *Mémoires sur les hôpitaux*, x.

36. *Extrait des Registres de l'Académie Royale des Sciences du 20 juin 1787, rapport des commissaires chargés, par l'Académie, des Projets relatifs de l'établissement des quatre Hôpitaux* (Paris: Imprimerie Royale, 1787). See also Tenon, *Mémoires sur les hôpitaux*, 349–472.

37. Tenon, *Mémoires sur les hôpitaux*, 351–4.

38. The second edition, published in 1768, was more widely available and has been referenced in this chapter. Théophile de Bordeu, *Recherches sur le pouls par rapport aux crises*, 3 vols. (Paris: Didot le jeune, 1768–1772). For background material on Bordeu and his contribution to French medicine in general see Elizabeth A. Williams, *The Physical and the Moral: Anthropology, Physiology and Philosophical Medicine in France 1750–1850* (Cambridge: Cambridge University Press, 1994), esp. chap. 1.

39. Xavier Bichat, *Recherches physiologiques sur la vie et sur la mort*, 3rd ed. (Paris: Brosson, 1805).

40. Edwin Clarke and L.C. Jacyna, *Nineteenth-Century Origins of Neuroscientific Concepts* (Berkeley, 1987). For a close reading of Bichat's work see Elizabeth A. Williams, *The Physical and the Moral: Anthropology, Physiology, and Philosophical Medicine in France 1750–1850* (Cambridge, 1994). The debate on sensualism is examined in Goldstein, *The Post-Revolutionary Self*.

41. Mathieu François Regis Buisson, *De la division la plus naturelle des phénomènes physiologiques considérés chez l'homme* (Paris, 1802).

42. "qui n'exerçant leur action que sur les molécules les plus ténues des corps, sont à portée d'en apprécier la nature intime, et servant à éclairer sur leur choix." Premier Extrait *"De la division la plus naturelle des phénomènes physiologiques considérés chez l'homme par M. Fr. R. Buisson,"* *Bibliothèque Médicale ou Recueil Périodique: D'extraits des meilleurs ouvrages de médecine et de chirurgie* 1, no. 1 (Paris, 1803), 26–44. Buisson's work was reviewed alongside Bichat's *Recherches physiologiques sur la vie et la mort*.

43. Second Extrait, Ibid., 160–82.

44. "diffère essentiellement de l'audition passive par cet acte de volonté qui en commande l'exercice, par l'attention qui l'accompagne, et c'est cette différence qui se trouve marquée dans le langage ordinaire par l'emploi qu'on y fait des mots *entendre* et *écouter*. On ne peut s'empêcher d'entendre, mais on n'écoute que parce qu'on le veut. On écoute afin de mieux entendre, et bien plus on peut quelquefois écouter sans entendre, lorsque, suivant l'expression vulgaire, on prête l'oreille pour reconnaître des sons dont on croyait avoir reçu une première impression." Ibid.

45. Jacalyn Duffin, *To See with A Better Eye: A Life of R.T.H. Laennec* (Princeton: Princeton University Press, 1998).

46. "Il est très vrai que la marche naturelle du pouls peut être comparée en général et en passant, aux accords qui résultent du mélange bien proportionné de plusieurs instruments de Musique; mais ce ne peut jamais être qu'une comparaison, qui n'a d'autre usage que de faire concevoir ce qu'il faut exprimer." Bordeu, *Recherches sur le pouls*, vol. 1, 132.

47. See the "Eloge historique de M. Marquet." In *Nouvelle méthode facile et curieuse pour connaître le pouls par les notes de la musique*, ed. François-Nicolas Marquet, 2nd ed. (Paris: Didot le Jeune, 1769), 203–11.
48. "Augmentée de plusieurs observations et Réflexions critiques, et d'une Dissertation en forme de Thèse sur cette Méthode; ou d'un Mémoire sur la manière de guérir la mélancholie par la Musique, &. de l'Eloge historique de M. Marquet, par M. Pierre-Joseph Buchoz." Ibid.
49. Ibid., 1.
50. "Tant et si longtemps que le mouvement du cœur et des artères est réglé, le corps de l'homme reste dans une santé parfaite; mais dès que ce mouvement se dérange par quelque accident, la santé se trouve altérée par une infinité de maladies." Ibid., 2–3.
51. "Réflexions de M. P. J. Buchoz, Docteur aggrégé au Collège Royal des Médécins de Nancy, sur la nouvelle Méthode de connaître le pouls par la Musique." Ibid., 166.
52. "On ne peut disconvenir que la Méthode que M. Marquet a donnée sur la connaissance du pouls, par la Musique, ne soit très ingénieuse et ne puisse servir à acquérir des lumières dans cette partie de la Sémiotique si nécessaire dans les diagnostics et pronostics." Ibid., 166–7.
53. "oscillations faibles et lentes, et épaissit les liquides par leurs stagnations; or, cet épaississement des liquides engendre la mélancolie." Ibid., 175–6.
54. "un ton égal et flexible." Ibid., 189.
55. "La Musique soit vocale, soit instrumentale, ou est diatonique, la plus ancienne de toutes, qui monte ou qui descend par différents tons; ou chromatique, qui ne diffère de la diatonique que par les semitons dont elle est ornée; ou enfin enharmonique, ornée de dièses et d'inflexions les plus douces des sens." Ibid., 190–1.
56. "La propagation du son se fait donc très vite, et parvient à l'instant à l'oreille, ensuite par un mécanisme admirable, dont a si bien parlé M. Duverney, il frappe le nerf auditif, par le moyen duquel il est porté jusqu'au *sensorium commune*, et là se forme l'idée du son." Ibid., 194.
57. "l'organe de l'ouïe est une espèce de tact; plus son choc est rude, plus il est offensé; plus il est doux, plus on ressent de plaisir." Ibid., 197.
58. "C'est en ce sens-là seulement qu'on peut dire avec Hérophile que les mouvements du pouls ont quelque rapport aux lois de la Musique; mais si on voulait appliquer au pouls les règles de la Musique, comme un Moderne l'a entrepris, on ne manquerait pas d'entrer dans des détails pénibles, qui n'en seraient pas pour cela plus utiles ni mieux fondés." Bordeu, *Recherches sur le pouls*, vol. 1, 132.
59. J. de Marque, "Rémarques preliminaires de l'éditeur." In *Recherches sur le pouls*, ed. Bordeu, vol. 3.
60. "Que n'aurais-je pas à dire de la fameuse circulation qui a tant ébloui, et qui est devenue chez les Mécaniciens, un instrument, dont ils se sont servis avec autant de confiance et de libéralité, que les Cartésiens en ont mis dans l'emploi de la matière subtile. Combien cette circulation a occasionné de mauvais raisonnements! Combien elle a rendu les Médecins inaccessibles aux bonnes et franches observations, faites sur les maladies et sur le corps vivant, qui formaient le fonds de la médicine ancienne!" Ibid., xiv.

61. "J'aimerais tant qu'un Musicien, pour me faire marcher sur les traces de Rameau, et m'apprendre à présider à un concert, formé d'un grand nombre d'instruments, s'amusât à me détailler, avec une trop savante profusion..." Ibid., xi.

62. "[A]idez-moi à saisir l'ensemble de tous ces divers sons variés, dont la combinaison fait le beau, le grand, le sublime de l'harmonie; montrez à mon oreille les moyens de saisir le plus léger ton, lorsqu'il passe ses bornes. Apprenez-moi l'histoire du corps vivant, dirais-je de même à un Physiologiste; nous avons tant analysé, tant et tant disséqué." Ibid., xii.

63. Jean-Joseph Menuret, *Nouveau traité du pouls* (Paris: Vincent, 1768).

64. "n'est qu'un mélange absurde et singulier de quelques dogmes Galénistes, des Mécaniciens et des Chimistes." Ibid., 114.

65. "Son ouvrage aurait été sûrement très avantageux, si le système, qui en fait la base, eût été moins conforme à celui des Mécaniciens, moins raisonné, et en un mot, plus rapproché de l'observation." Ibid., 118.

66. For Bordeu's physical methods of pulse-taking (points on the body) see Bordeu, *Recherches sur le pouls*, vol. 1, 337–56.

67. For a recent discussion on Chinese and Greek methods of pulse diagnosis see Shigehisa Kuriyama, *The Expressiveness of the Body and the Divergence of Greek and Chinese Medicine*, illustrated ed. (New York: Zone Books, 1999).

68. "Il [Galien] prétendit avoir trouvé des pouls qui ressemblaient à la marche des fourmis, il les appela *formicans*; d'autres qui allaient en diminuant [comme] la queue d'un rat, il les nomma *miures*; et il appela, d'après Hérophile, pouls *caprizans*, ceux qu'il crut représenter les sauts d'une chèvre." Bordeu, *Recherches sur le pouls*, vol. 1, xvii.

69. "Galien en faisant son Traité du pouls, raisonna beaucoup plus qu'il n'avait observé: il comprit pourtant que les différentes espèces de pouls devaient être distribuées en plusieurs classes: mais il y avait de la difficulté à les caractériser, à les rendre reconnaissables, et encore à les exprimer d'une manière assez intelligible." Ibid., xvi–xvii.

70. "Les Modernes s'en sont tenus à des divisions et à des dénominations plus simples, même en apparence plus significatives: on a divisé les pouls en *forts et faibles, fréquents et lents, grands et petits, durs et mous, etc.*" Ibid., xix.

71. "Mais il est facile d'apercevoir que cette *nomenclature* adoptée par les Modernes, a presqu'autant de défauts que celle qu'ils ont rejetée, parce que dans le fait ces dénominations n'expriment rien d'assez précis; il n'est pas possible de déterminer quel signe on doit juger dans les maladies que le pouls est par exemple *dur* ou *mou*, *grand* ou *petit*; sa *petitesse* et sa *grandeur*, sa *mollesse* et sa *dureté*... d'ailleurs il n'arrive que trop souvent qu'un pouls qui est trouvé *grand* ou *dur* par un Médecin, paraîtra *petit* ou *mou* à un autre: ainsi ces définitions ou ces dénominations ne peuvent rien d'exprimer d'assez positif." Ibid., xix–xx.

72. "Les maladies dont les crises sont précédées et annoncées par des pouls *simples*, ne sont jamais des maladies de mauvaise espérance; celles au contraire dans lesquelles se trouvent les pouls *compliqués*, sont ordinairement des maladies graves: or, comme il s'en faut de beaucoup que les différents ressorts du jeu de l'économie animale se rendent aussi sensibles, aussi reconnaissables dans de médiocres lésions des fonctions que dans un état de grande maladie; ce n'est donc que dans l'exposition des pouls compliqués qu'on a dû placer

les examens et les discussions qui ont conduit aux principes féconds et aux importantes règles qu'on a cherché à établir." Ibid., xxx.

73. "beaucoup plus loin: & les ramener par-là à des principes généraux propres à répandre sur la théorie de l'art, autant de lumière que sur la pratique." Ibid., xxviii.

74. "*La dureté, la mollesse, la grandeur, la fréquence,* etc. ne sont que *des états, des modes relatifs* qui ne peuvent être évalués que par une mesure commune et fixe, à laquelle on puisse rapporter toutes ces variations." Ibid., 2–3.

75. "Le pouls naturel des vieillards est beaucoup plus *fort*, beaucoup plus *dilaté*, beaucoup plus *dur* que celui des enfants." Ibid., 5.

76. "Le pouls naturel des femmes est, en général, plus *vif* et plus approchant de celui des enfants..." Ibid., 7.

77. Ibid., 13.

78. "Il y a, par exemple, des pouls qui seront appelés, *petits, serrés, durs, pleins, dilatés, développés*: c'est comme si on disait qu'ils sont plus *petits*, plus *pleins*, plus *mous*, pouls *développés*, que dans l'état ordinaire ou naturel du sujet qu'on examine." Ibid., 12.

79. "il faut, à l'exemple de tous les Médecins, rassembler, lorsqu'on juge de l'état d'une maladie, tous les symptômes, et peser toutes les circonstances: dans combien d'écueils ne tomberait-on pas sans cette précaution?" Ibid., 15.

80. "le pouls naturel et parfait des adultes." Ibid., 16.

81. "égal, ses pulsations se ressemblent parfaitement, elles sont à des distances parfaitement égales: il est mollet, souple, libre, point fréquent, point lent, vigoureux, sans paraître faire aucune sorte d'effort." Ibid., 16.

82. Ibid., 18–19.

83. "mou, plein, dilaté, ses pulsations sont égales; on sent dans chacune une espèce d'ondulation; c'est à dire que la dilation de l'artère se fait en deux fois; mais avec une aisance, une mollesse et une douce force d'oscillations." Ibid., 30–1.

84. "fort, avec un redoublement dans chaque battement, il est moins mou, moins plein, souvent plus fréquent que le pouls pectoral; il paraît être intermédiaire entre le pouls pectoral décrit dans le Chapitre précédent et le nasal qui sera décrit dans le Chapitre suivant..." Ibid., 42.

85. Ibid., 4.

86. "Les Médecins les plus clairvoyants et les plus assurés sur ce genre de connaissances, sont ceux dont la tête est la mieux fournie de toutes les images des différentes espèces de pouls." Ibid., 7.

87. See Observation VI, Ibid., 36–7.

88. See Observation IX, Ibid., 38.

## 4 The Blind and the Communication-Object

1. *Grand concert extraordinaire exécuté par un détachement des quinze-Vingts au Caffé des Aveugles, Foire Saint Ovide au Mois de Septembre 1771, estampe* (Paris: Chez Mondhare rue St Jacques, n.d.). BNF EP.

2. *Troisième note du citoyen Haüy, auteur de la manière d'instruire les aveugles ou court exposé de la naissance des progrès et de l'état actuel de l'Institut national des aveugles-travailleurs au 19 brumaire an IX de la République française*

[10 Nov. 1800] *entremêlée de quelques observations relatives à cet établissement* (Paris, 1800, 9. Musée Valentin Haüy).

3. "Qu'il nous soit permis de rendre hommage aux talents et au zèle de M. l'Abbé de l'Epée qui a ouvert la carrière de l'instruction aux Sourds & Muets, M. Haüy devient à son exemple le bienfaiteur des aveugles, et cette partie souffrante de l'humanité lui devra les moyens de bonheur que l'on ne croyait pas pouvoir espérer pour elle." "Rapport de l'Académie Royale des Sciences, 18 Feb. 1785." In Valentin Haüy, *Essai sur l'éducation des aveugles* (Paris: Imprimé par les Enfans-Aveugles, 1786), 12.

4. Rosenfeld, *A Revolution in Language*, 123.

5. Jean-Philippe Rameau, *Code de musique pratique, ou méthodes pour apprendre la Musique même à des Aveugles [...]* (Paris: Imprimerie Royale, 1761). See the discussion of Rameau's *Code* by Robert W. Wason, "*Musica Practica*: Music Theory as Pedagogy." In *The Cambridge History of Western Music Theory*, ed. Thomas Christensen (Cambridge: Cambridge University Press, 2002), 46–77.

6. Etienne Bonnot de Condillac, *Essai sur l'origine des connaissances humaines, ouvrage où l'on réduit à un seul principe tout ce qui concerne l'entendement humain*, 2 vols. (Amsterdam: Mortier, 1746). The translations are from Condillac, *An Essay on the Origin of Human Knowledge Being a Supplement to Mr Locke's Essay on Human Understanding translated from the French by Mr Nugent* (London: Nourse, 1756).

7. Pierre Henri, *La vie et l'œuvre de Valentin Haüy* (Paris: Presses Universitaires de France, 1984), 24–32.

8. Ibid., 21.

9. Haüy, *Adresse du citoyen Haüy, auteur des moyens d'éducation des enfans aveugles et leur premier instituteur aux 48 sections de Paris, présentée à la suite d'une adresse de la section de l'Arsenal, en date de l'an I de la République française le 13 décembre 1792, dont il étoit porteur.*

10. *Troisième note du citoyen Haüy.*

11. "Les Enfants Aveugles vous invitent par ma bouche à venir." *Adresse du citoyen Haüy.*

12. Haüy, *Essai.*

13. Ibid., 83.

14. "semblaient être devenus tout Oreilles." *Troisième note du citoyen Haüy.*

15. "Personne n'ignore la délicatesse de ce sens chez des individus, qui, depuis l'enfance, s'en servent pour remplacer celui que la Nature leur a refusé." Haüy, *Essai*, 26–7.

16. The title page includes the note that the book was "Imprimé par les Enfans-Aveugles, sous la direction de M. Clousier, Imprimeur du Roi."

17. These processes are described in detail in Haüy's *Essai* but also in Henri, *La vie et l'œuvre.*

18. "Nous ne prétendons pas mettre jamais le plus habile de nos Aveugles en concurrence dans aucun genre, même avec le plus médiocre des Savants ou des Artistes clairvoyants." Ibid., 11–12.

19. Ibid., 34–6.

20. Ibid., 36.

21. Epreuves des caractères des aveugles gravés et fondus pour être imprimés en relief. Paris, n.d.

22. Ibid., 85–6.

23. Ibid., 83–5.
24. "La Lecture est le vrai moyen d'orner la mémoire d'une manière facile et prompte...Sans elle, les productions littéraires ne formeraient dans l'esprit humain qu'un amas désordonné de notions vagues." Ibid., 15–16.
25. "A notre tour permettez-nous de vous interroger. Que sert-il que l'on imprime des livres chez tous les peuples qui vous environnent? Lisez-vous le Chinois, Le Malabar, le Jurc, les Quipos des Péruviens, et tant d'autres langages si nécessaires, à ceux qui bien entendent? Eh bien! vous ne seriez qu'un aveugle en Chine, sur les rives du Gange, dans l'Empire Ottoman, au Pérou." Ibid., 37.
26. "Nous en appelons à vous enfin tendres et respectables époux! nés dans le sein d'une fortune honnête; vous dont le fils vient de naître, et cependant ne verra jamais le jour; quelle douce satisfaction pour nous de pouvoir modérer les transports de votre douleur." Ibid., 40–41.
27. For details of these performances see Henri, *La vie et l'œuvre*, 77–85, and Zina Weygand, *The Blind in French Society: From the Middle Ages to the Century of Louis Braille*, trans. Emily-Jane Cohen with a preface by Alain Corbin (Stanford: Stanford University Press, 2009), 105–10, 146–52.
28. Bernard Lamy, *La Rhétorique ou l'art de parler*, 4th ed. (Paris: Pierre Debats and Imbert Debats, 1701 [1676]). All my translations are from *The Art of Speaking Rendered into English* (London: Godbid, 1676).
29. Rosenfeld, *A Revolution in Language*, 33.
30. "faux éclat." Bernard Lamy, *Nouvelles Réflexions sur l'Art Poetique* (Paris, 1768), 23.
31. Lamy's work was very influential in music-making. See Jonathan Gibson, "'A Kind of Eloquence Even in Music': Embracing Different Rhetorics in Late Seventeenth-Century France," *The Journal of Musicology*, 25, no. 4 (Fall 2008), 394–433.
32. B. Lamy, *Nouvelles Réflexions sur l'Art Poetique*.
33. "Les Poètes s'étudièrent peu à peu à composer leur Ouvrages selon le goût de leurs Auditeurs, dont le plaisir fut la seule règle qu'ils suivirent dans la conduite de leurs ouvrages." Ibid., iii.
34. "la peinture parlante." Ibid., 1.
35. "Les hommes ne voyent pas en plus, que Dieu est le principe et le terme de ce mouvement ou de cette inclination de leur cœur, qui leur fait aimer la grandeur, et rechercher la béatitude dans l'état où ils sont." Ibid., 6.
36. See in particular chapters 4–8 of Lamy's text.
37. "la beauté des créatures" created a "faux éclat" which was so loud that "la voix de la nature, qui crie sans cesse." Ibid., 17.
38. Ibid., 26.
39. "Il est convaincu de la vanité des créatures, et qu'elles ne lui peuvent procurer cette félicité par les forces qu'il souhaite: il sait aussi qu'il ne peut acquérir cette félicité par les forces qu'il trouve en lui-même. Il voit sa faiblesse, mais il ne cherche point le secours qui lui est nécessaire, il se sent enveloppé d'épaisses, mais il ne demande point de flambeau pour les dissiper." Ibid., 31.
40. Lamy, *The Art of Speaking*, 35.
41. Ibid., 64.
42. Ibid., 4.
43. Ibid., 113.

44. Ibid., 101.
45. Ibid., 146–7.
46. Ibid.
47. Ibid., 115.
48. Ibid., 116.
49. Ibid., 117–18.
50. Ibid., 128.
51. Ibid., 133.
52. Ibid., 136.
53. Ibid., 108.
54. Ibid., 115.
55. Ibid., 125.
56. Ibid., 148.
57. Condillac, *An Essay on the Origin of Human Knowledge*, 173.
58. Condillac, 181.
59. Ibid., 181.
60. Ibid., 225.
61. "L'oreille est comme le vestibule de l'âme. Si vous blessez l'oreille par un son désagréable, l'âme sera mal disposée à recevoir ce que vous lui présentez." Crevier, *Rhétorique française*, vol. 1 (Paris: Saillant, 1767), 4.
62. Ibid., xv.
63. Gabriel-Henri Gaillard, *Rhétorique française à l'usage des jeunes demoiselles avec des exemples tirés pour la plupart de nos meilleurs Orateurs & Poètes Modernes*, 5th ed. (Paris: Veuve Savoie, 1776).
64. "En effet, l'Orateur, quelque sujet qu'il traite, a dans les Auditeurs des Juges à prévenir a persuader; il a, dans l'Objet de son discours, un Client à defendre, une cause à plaider, une Proposition à établir nettement et à prouver solidement; il a enfin une récapitulation courte, véhémente et rapide à faire de ses plus fortes preuves." Ibid., xii.
65. "Pour pouvoir tous ces grands effets, il faut commencer par plaire; c'est le puissant ressort qui fait mouvoir toute la machine de l'esprit et du Cœur humain." Ibid., 14.
66. Rameau, *Code de musique*, xiii.
67. Ibid., 139, 167.
68. Dora B. Weiner, *The Citizen-Patient*, 232–6; Weygand, *The Blind in French Society*, 122–35.
69. See Weygand, *The Blind in French Society*, 113.
70. Henri, *La vie et l'œuvre*, 106–17.
71. Ibid., 240.
72. Henri, *La vie et l'œuvre*, 93–8.
73. "répandent dans le public les bruits les plus calomnieux, prétendant que les patriotes vont être arrêtés, que déjà plusieurs d'entre eux le sont, que leur but, en semant ces faux bruits, est de glacer le courage des bons citoyens et de les rallier autour de leur parti de leur sauver." "Sur l'observation faite par un de ses membres, 28 ventose an II," Item 2019, In Alexandre Tuetey, *Publications relatives à la Révolution française*, vol. 10, (Paris: Imprimerie nouvelle, 1890–1914), 451.
74. See for example, "Décret de la Convention nationale, 20 juil. 1793," Item 2996, In Tuetey, *Publications Relatives*, vol. 8, 484.

75. See for example, "Déclaration du sieur Cornet...au sujet de bruits souterrains entendus pendant la nuit, 20 juillet 1792," Item 227, In *Publications Relatives*, vol. 6, 30.
76. See Lynn Hunt's discussion of Mona Ozouf's work in Lynn Hunt, *Politics, Culture and Class in the French Revolution* (Berkeley: University of Los Angeles, 1984), 23.
77. Haüy, *Adresse du citoyen Haüy*.
78. Petition à l'Assemblée Nationale par des Aveugles suivants l'Institution de M. Haüy en 1791, C167, no. 149, Archives parlementaires, t. XLVI. See also the Procès-Verbaux de Comité des secours publics, AF*II 39, folio 11. AN.
79. "connaissent bien les faiblesses de ses moyens, vit que de telles vérités mises au jour éclipseraient nécessairement sa gloire, et renverseraient ses projets ambitieux." Ibid.
80. "Nous serve à atteindre une loi générale sur tous les aveugles." Ibid.
81. "Pénétrez avec nous dans cet établissement; nous serons les guider les plus sûrs pour vous conduire dans cet antre du charlatanisme. Nous espérons Législature, que les Motifs qui nous ont mérité la haine de M. Haüy, pourront nous mériter la bienveillance de l'Assemblée Nationale. Nous vous supplions d'acquiescer à notre demande, afin que la vérité Triomphe de l'imposture, et que les Aveugles ne soient plus désormais les victimes de la Cupidité des hommes." Ibid.
82. See Ingrid Sykes, "Sounding the 'Citizen-Patient': The Politics of Voice at the Hospice des Quinze-Vingts in Post-Revolutionary Paris," *Medical History* 55, vol. 4 (2011), 479–502.
83. "L'humanité, la justice et l'intérêt general sollicitent la conservation de ce précieux hospice." *Observations pour les aveugles de l'hôpital des Quinze-Vingts. Sur le projet de décret du comité de secours de la convention nationale, pour la suppression de cet hospital* (Paris, 179–), 21.
84. Note, "difficultés d'ouvrages," *Musée de Valentin Haüy*.
85. "Petition à l'Assemblée Nationale." AN.
86. See Haüy's Réponse à diverses objections contre la Lecture à l'usage des Aveugles, in his *Essai*, 26–32.
87. Henri, *La vie et l'œuvre*, 106–17.
88. "le nommé Haüy, ancien interprète de tyran et de l'Amirauté, instituteur des Aveugles, intriguant et faux patriote, qui a cherché à faire de son établissement un foyer de fanatisme, ayant fait afficher dans tout Paris qu'on pouvait aller y entendre la messe." *Déclaration de membres des Comités civil et revolutionnair*, 24 ventose an II, Item 165, Tuetey, *Publications Relatives*, vol. 11, 99.
89. "Je fus trois fois incarcéré comme terroriste." Henri, *La vie et l'œuvre*, 106.
90. On Chaptal's workers ethic for the blind see Weygand, *The Blind in French Society*, 240–52.
91. Ibid.
92. See "Etablissement National des Quinze-Vingts: Reglèment de Police de l'Intérieur de l'établissement," 13 June 1800, F15 2581, AN.
93. He did not return to France until just before his death in 1817.
94. L'Abbé Haüy, *Traité élémentaire de physique médicale* (Paris: Delahays, 1855) 85–115.

95. "mouvement imprimé par la percussion, ou de toute autre manière, aux moléculesd'un corps." Ibid., 85–6.
96. Valentin Haüy, *Nouveau syllabaire…manuel de l'élève [et manuel de l'instituteur], par le citoyen Haüy* (Paris: Institut national des aveugles travailleurs, an VIII [1799–1800]).
97. "Nous avons dans notre langue une quantité de sons simples dont nous ne trouvons point les caractères dans notre alphabet." Ibid.
98. Weygand, *The Blind in French Society*, 261–91.

## 5   Sound, Health and the Auditory Body-Politic

1. "Ils ont reçu cette année, pour la première fois, des leçons de harpe. On n'avait pas cru possible, jusqu'à présent, de leur apprendre cet instrument si difficile pour les clairvoyants eux-mêmes, par la position pénible du corps et la multiplicité des cordes que rien ne différencie." Sebastian Guillié, *Essai sur l'instruction des aveugles*, 3rd ed. (Paris: Imprimé par les aveugles, 1820), 215.
2. For information on the programme see Sykes, *Women, Science and Sound*, chap. 4.
3. "Ce son est-il un corps réellement existant dans la nature, et qui viendrait frapper l'oreille? ou bien n'est ce qu'une manière dont le corps en mouvement pouvant nous affecter? et dans ce cas, le son ne serait-il que l'exercice de l'oreille appliqué à la connaissance du mouvement des corps." Jean-Joseph Pascal, *De l'ouïe considérée dans ses rapports avec l'intelligence et la voix* (Paris: Didot le Jeune, 1821), 13.
4. Some unique models remain in the Musée de la musique, Paris.
5. Etienne-Louis Geoffroy, *Dissertations sur l'organe de l'ouïe, 1. De l'homme, 2. Des reptiles, 3. Des Poissons* (Amsterdam, Paris: Cavelier, 1778).
6. See Emma Spary, *Utopia's Garden: French Natural History from the Old Regime to the Revolution* (Chicago: Chicago University Press, 2000), 197–8.
7. *Histoire de la Société Royale de Médecine*, année 1776 (Paris: Philippe-Denys Pierres, 1779), 95–7.
8. "destiné uniquement à le recevoir et à en communiquer l'impression au cerveau, par le moyen du nerf auditif." Geoffroy, *Dissertations*, 30.
9. "doit être plus forte et plus considérable." Ibid., 39.
10. "L'organe de l'ouïe est le seul de tous les sens, qui peut mesurer exactement son objet. Les yeux distinguent bien les couleurs, mais ils ne peuvent établir des degrés fixes entre leurs nuances: il en est de même de l'odorat et du tact. L'ouïe seule, non seulement distingue les sons, mais les mesure avec tant d'exactitude que l'on est parvenu à distinguer les tons, les demi-tons, et leurs différentes modifications. C'est de là qu'est né l'art du chant, qui est si naturel à l'homme, que tous les peuples ont eu leur musique, et que l'harmonie flatte tous les Auditeurs, sans qu'ils soient Musiciens. Sans doute que si la rétine eût été partagée, comme le nerf auditif, en petites cordes ou fibrilles de différente longueur, l'œil pourrait aussi bien mesurer la lumière, que l'oreille mesure et distingue les sons." Ibid., 38–9.
11. See Ingrid Sykes, "Sounding the Citizen Patient."

12. Sean Quinlan, "Physical and Moral Regeneration after the Terror: Medical Culture, Sensibility and Family Politics in France, 1794–1804," *Social History* 29, no. 2 (May 2004), 139–64.
13. Ibid., 143.
14. Ibid., 140.
15. Marc-Antoine Petit, *Essai sur la médecine du coeur* (Lyon: Garnier, 1806), 138.
16. "toutes les passions sont en jeu; toutes les âmes sont exaltées: la sensibilité double les forces; l'énergie est partout; et tout homme s'indigne à la seule idée d'une injustice." Ibid., 119.
17. "Une grande nation qu'on veut régénérer est comme un orgue qu'on veut remonter. L'artiste ne brise pas chaque tuyau qui rend les sons faux ou discordants: il le met au ton qu'il désire, et, quand il touche le premier air, il enchante ses auditeurs." *Journal de Rouen*, le 28 pluviose an IV In *Musique et Révolution française*, ed. Michelle Biget (Paris: Les Belles Lettres, 1989), 221.
18. Quinlan, "Physical and Moral Regeneration," 141.
19. Joseph-Louis Roger, *Traité des effets de la musique sur le corps humain* edited and translated from the Latin text by Etienne Sainte-Marie (Paris: Brunot, 1803). See Etienne Sainte-Marie's later work on natural medicine, *Nouveau formulaire médical et pharmaceutique* (Paris: Rey et Gravier, 1820).
20. "Ceux qui regardent la musique comme un art purement agréable, ne croiront pas aux effets avantageux que l'auteur de cet ouvrage lui attribue dans un grand nombre de maladies." Sainte-Marie in Roger, *Traité des effets*, vii.
21. "se répand dans tout le corps, et le dilate … et y réveille le courage, l'amour, la bienfaisance, la pitié, la joie, les passions expansives." Ibid., x–xi.
22. "La santé n'est pas l'état le plus favorable aux effets de la musique. Cet art agit plus vivement sur un corps affaibli par quelque maladie." Ibid., xxii.
23. "qui constituent dans les organes la vie elle-même." Ibid., xiv.
24. "la musique est un exercice, et on doit la recommander aux femmes et aux gens des lettres, qui mènent une vie sédentaire." Ibid., xiii.
25. "On observe parmi les musiciens plus de vieillards que parmi les autres artistes." Ibid., xvii–xviii
26. Roger, *Traité des effets*, 252–3.
27. V. Forgues, *De l'influence de la musique sur l'économie animale* (Montpellier: Martel, 1802).
28. See for example, Desessartz, *Résultats des observations faites dans plusieurs départements de la République, sur les maladies qui ont régné pendant les six premiers mois de l'An VIII* (Paris, Huzard, An VIII).
29. "Pourquoi l'emploi de cet art, qui produit des effets si étonnants dans l'homme en santé et dans l'homme malade, est-il abandonné aujourd'hui par les médecins? Pourquoi n'y aurait-on pas encore recours dans des maladies qui éludent l'action des médicaments vantés comme les plus énergiques?" Jean-Louis Desessartz, *Réflexion sur la musique considérée comme moyen curatif* (Paris, 1802), 3.
30. Ibid., 3.
31. "Ainsi, tout est matière, et matière en action: ce n'est point une qualité occulte." Ibid., 19.

32. Pascal, *De l'ouïe considérée*, 32–4.

33. Jules-Charles Teule, *De l'oreille: essai d'anatomie et de physiologie, précédé d'un exposé des lois de l'acoustique* (Paris: Gabon, 1828).

34. "On admettra facilement aussi que la présence d'un organe spécial de l'audition dans une espèce suppose un chant, un cri..., comme moyen d'établir les rapports nécessaires de l'individu à l'espèce." Ibid., 219.

35. "il suffit pour s'en convaincre d'observer, qu'eu égard à la vélocité du son, on peut admettre que la perception a lieu rigoureusement au même instant pour chaque oreille, et que les impressions transmises à la cause du sentiment par le double organe de l'ouïe, sont identiques ou seulement différentes dans leur intensité, ce qui n'empêche pas de les confondre." Ibid., 221.

36. "la diversité des directions de l'ébranlement communiqué à la pulpe nerveuse auditive. Elle est un produit de l'éducation; il est indispensable, en effet que, par des comparaisons multipliées, l'esprit apprenne à saisir la liaison qui existe entre les positions relatives du corps sonore et de l'oreille, et de la diversité des sensations de l'ouïe." Ibid., 223.

37. "Il semble que le sentiment de la mesure dérive de l'organisation même où l'on voit les mouvements de la circulation assujettis à l'emprise du rythme, et par suite tous les organes excités par le sang d'une manière périodique et régulière." Ibid., 232.

38. "Les aveugles de naissance portent généralement un jugement plus certain que les autres hommes sur les objets dont on acquiert la connaissance par l'oreille; il n'est pas rare qu'un aveugle reconnaisse au bruit de la marche les personnes dont il était séparé depuis longtemps." Ibid., 236.

39. *Description des machines et procédés specifiés dans les brevets d'invention, de perfectionnement et d'importation*, vols. II–XXVIII (Paris: Madame Huzard/Imprimerie Nationale, 1818–1837).

40. "Brevet d'invention 919, 13 juillet 1809, Pour une nouvelle harpe sur laquelle on peut faire les dièses et les bémols, Aux sieurs Erard frères, à Paris." In *Description des machines et procédés specifiés dans les brevets d'invention, de perfectionnement et d'importation* X (Paris: Madame Huzard/Imprimerie Nationale, 1825), 287.

41. "plus pur et plus intense." "Brevet d'invention 64, 17 mars 1799, Pour un mécanisme particulier, destiné à tendre les cordes de harpe, aux sieurs Ruelle et Cousineau, père et fils, à Paris." In *Description des machines* II (Paris, 1818), 37–41.

42. "le mécanisme est indestructible... s'il se trouve dans les mains exercées." "Brevet d'invention 125, 20 juillet 1802, Pour une nouvelle mécanique de harpe, à plans inclinés paraboliques, et à renfo cemens acoustiques, inventée par Michel-Joseph Ruelle, mais dont la jouissance a été cédée au sieur Cousineau." In *Description des machines* II (Paris, 1818), 169–72.

43. "Brevet d'invention 685, 7 novembre 1815, Pour une harpe d'harmonie, Au sieur Thory, à Paris." In *Description des machines* VIII (Paris, 1824), 366.

44. "Brevet d'invention 344, 9 juin 1809, Pour une harpe-harmonico-forté au sieur Keyser de l'Isle, à Paris." In *Description des machines* V (Paris, 1823), 85–6.

45. "Brevet d'invention 898, 11 juin 1818, Pour un perfectionnement ajouté à la harpe, Au sieur Mérimée." In *Description des machines* X (Paris, 1825),

227–9; "Brevet d'invention 912, 24 novembre 1813, Pour une nouvelle mécanique de harpe, Au sieur Plane, à Paris." In *Description des machines* X (Paris, 1825), 268–9; "Brevet d'importation 1690, 7 décembre 1814, Pour un mécanisme applicable aux harpes ordinaries, Au sieur Gilles, à Paris." In *Description des machines* XIX (Paris, 1830), 124–6.

46. "Brevet d'invention 144, 13 juillet 1803, Pour un instrument nommé piano-harmonica, au sieur Tobias Schmidt, facteur de pianos, à Paris." In *Description des machines* II (Paris, 1818), 242–50.

47. "Brevet d'invention 2088, 23 mars 1827, Pour un piano à clavier placé sur les cordes, Au sieur Côte, facteur de pianos, à Lyon, département du Rhône." In *Description des machines* XXIII (Paris, 1832), 245–8.

48. "Brevet d'invention 652, 8 septembre 1815, Pour des piano-fortés carrés à six octaves, cinq pédales et tambour guerrier, Au sieur Thory, à Paris." In *Description des machines* VIII (Paris, 1824), 219; "Brevet d'invention 975, 20 octobre 1809, Pour la construction d'un forté-piano en forme de clavecin, Aux sieurs Erard frères, fabricans de forté-pianos, à Paris." In *Description des machines* XI (Paris, 1825), 70–1; "Brevet d'invention 2238, 30 octobre 1827, Pour un piano qui diffère des pianos connus par la position des chevilles et des étouffoirs, Au sieur Cluesman (Jean-Baptiste), facteur de pianos, à Paris." In *Description des machines* XXV (Paris, 1834), 17; "Brevet d'invention 2790, 2 février 1827, Pour un piano de forme et de construction nouvelles, muni d'un nouveau mécanisme, Au sieur Dietz fils, facteur de pianos, à Paris." In *Description des machines* XXIX (Paris, 1836), 331–2.

49. "Brevet d'invention 2434, 13 septembre 1828, Pour un sommier prolongé s'adaptant également aux pianos carrés et à queue, Aux sieurs Pleyel (Ignace) et compagnie, fabricans de pianos, à Paris." In *Description des machines* XXVII (Paris, 1835), 41–3; "Brevet d'invention 2884, 28 septembre 1827, Pour un piano à sommier isolé, Au sieur Triquet, facteur de pianos, à Paris." In *Description des machines* XXX (Paris, 1836), 203–4.

50. "Brevet de perfectionnement 1665, 31 janvier 1824, Pour un nouveau mécanisme à échappement qui s'adapte aux forté-pianos, Au sieur Klepfer-Dufaut (Henri), facteur de pianos, à Lyon." In *Description des machines* XIX (Paris, 1830), 5–7; "Brevet d'invention 2478, 10 février 1829, Pour un nouvel échappement applicable aux pianos droits et aux pianos verticaux, Aux sieurs Blanchet et Roller, fabricans de pianos, à Paris." In *Description des machines* XXVII (Paris, 1835), 188–90.

51. "Brevet d'invention 1333, 31 janvier 1812, Pour un forté-piano d'une forme et d'un mécanisme particuliers, à sieur Erard frères, à Paris." In *Description des machines* XIV (Paris, 1827), 295–7; "Brevet d'invention 1332, 31 janvier 1812, Pour un forté-piano ayant la forme d'un sécretaire, Aux sieurs Erard frères, à Paris." In *Description des machines* XIV (Paris, 1827), 292–4.

52. "Brevet d'invention 2866, 2 juin 1825, Pour un piano de forme elliptique et à deux tables d'harmonie, Au sieur Eulriot (Etienne), facteur de pianos, à Paris." In *Description des machines* XXX (Paris, 1836), 144–6.

53. "Brevet d'invention 1380, 15 novembre 1820, Pour des procédés de construction d'un nouveau piano, Au sieur Wagner (Jean-Baptiste), luthier, facteur de pianos, à Arras, département du Pas-de-Calais." In *Description des machines* XV (Paris, 1828), 61–6.

54. "Brevet d'invention 1808, 17 novembre 1825, Pour un piano à une corde, appelé piano unicorde, Aux sieurs Pleyel, père et fils aîné." In *Description des machines* XXI (Paris, 1831), 65–7.

55. "Brevet d'invention 1334, 4 février 1812, Pour un forté-piano à son continu, Aux sieurs Erard frères, à Paris." In *Description des machines* XIV (Paris, 1827), 298; "Brevet d'invention 1602, 10 fèvrier 1819, Pour un appareil appelé sostenente, Aux sieurs Mott (Julius-César), Mott (Isaac-Henri-Robert), et compagnie." *Description des machines* XVIII (Paris, 1829), 18–24.

56. "Brevet d'importation 1177, 31 décembre 1810, Pour des procédés de fabrication de cordes en cuivre et en fer, à l'usage du forté-piano et autres instrumens de musique, Par M. Pleyel (Ignace), à Paris." In *Description des machines* XIII (Paris, 1827), 49–56; "Brevet d'invention 2050, 9 mars 1827, Pour une nouvelle méthode de faire des cordes harmoniques sans noeuds et d'une seule longeur pour chaque instrument, Aux sieurs Savaresse et Compagnie, fabricans de cordes harmoniques, plaine de Grenelle, près Paris." In *Description des machines* XXIII (Paris, 1832), 138–41. "Brevet d'invention 1259, 25 avril 1822, Pour un mécanisme propre à fixer les chevilles des instruments de musique à corde, nommé fixateur, Au sieur Legros de la Neuville, à Paris." In *Description des machines* XIV (Paris, 1827), 41–2; "Brevet d'invention 1428, 12 juin 1823, Pour une cheville à frein, dont l'effet est de maintenir, dans leur accord et dans leur tension, les cordes de violins, basses et guitarres, Aux sieurs Brouet (Henri-Victor), mécanicien Clément (Jacob), luthier, tous deux à Paris. In *Description des machines* XV (Paris, 1828), 315.

57. "Brevet d'invention 1391, 11 décembre 1817, Pour des procédés de construction des instrumens de musique à cordes et à archet, tels que le violon, l'alto, la basse et la contre-basse, Au sieur Chanot (François), officier du génie maritime, à Paris." In *Description des machines* XV (Paris, 1828), 161–79.

58. "Brevet de perfectionnement 466, 19 septembre 1806, Pour un instrument de musique imitant le forté-piano appelé harmomelo, Au sieur Pfeiffer et compagnie, à Paris." In *Description des machines* VI (Paris, 1824), 266–70.

59. "Depuis longtemps on a cherché à composer un instrument à cordes dans une position perpendiculaire, pour le jouer en clavier, et toujours on a trouvé que les cordes tendues en cette position produisent un son beaucoup plus fort, plus plein, et plus agréable que celles tendues horizontalement." Ibid., 266.

60. "Brevet d'invention 964, 18 février 1814, Pour un instrument de musique appelé clavi-harpe, Aux sieurs Dietz et Second." In *Description des machines* XI (Paris, 1825), 15–17.

61. "Le clavier de cet instrument produit, au moyen du mécanisme qui y correspond, absolument le même effet que les doigts en pinçant de la harpe." Ibid., 17.

62. "Brevet d'importation 1799, 6 octobre 1825, Pour un instrument de musique portativ nommé guitare-harpe, Au sieur Levien (Mordaunt), professeur de musique à Londres." In *Description des machines* XXI (Paris, 1831), 43–45; "Brevet d'invention 464, 13 août 1811, Pour une guitare-lyre, qui présente plusieurs avantages sur les guitares faites jusqu'à ce jour, Au

sieur Mougnet, luthier, à Lyon, départmenet du Rhône." In *Description des machines* VI (Paris, 1824), 263.

63. "J'obtiens un son qui, au lieu d'être trop en dehors, comme on le reproche à la guitare, et de sortir avec confusion, devient, en vibrant, plus rond, plus harmonieux et plus rapproché de la beauté du son des meilleures harpes, par le placement analogue des veines de sapin." "Brevet d'invention 1962, 15 décembre 1826, Pour une guitare a dix cordes appelée décacorde, Aux sieurs Lacote, luthier et Carully, compositeur de musique." In *Description des machines* XXII (Paris, 1832), 258–9; "Brevet d'invention 2426, 28 novembre 1829, Pour une guitare perfectionnée, dans laquelle les veines de la table sont en travers, Au sieur de Lacoux, à Paris." In *Description des machines* XXVI (Paris, 1835), 377.

64. "Brevet d'invention 237, 21 novembre 1806, Pour un instrument de musique dans le genre de la lyre-guitare, appelé lyre organisée, Au sieur Led'huy, à Coucy-les-Châteaux." In *Description des machines* IV (Paris, 1820), 45–7.

65. "Quant au *toucher*, il exige une grande habitude et une main bien exercée pour pouvoir obtenir du jeu du clavier ces sons doux et veloutés qui plaisent à l'oreille délicate, et qui sont si propres à émouvoir." Ibid., 47.

66. "Brevet d'invention 2928, 28 fèvrier 1831, Pour un nouveau système de guitare appelé guitarion, Au sieur Franck (Maximilien), artiste, à Paris." In *Description des machines* XXX (Paris, 1836), 372–73.

67. "renferme un mécanisme au moyen duquel on obtient le *crescendo*, le *diminuendo* et le *tremolo*, plus un effet de grosse caisse, au moyen de quatre pédales." "Brevet d'invention 2483, 19 mars 1829, Pour un instrument de musique appelé harpo-lyre, Au sieur Salomon (Jean-François), professeur de musique à Besançon, département du Doubs." In *Description des machines* XXVII (Paris, 1835), 201–4.

68. "Brevet d'invention 1222, 13 décembre 1821, Pour un cor d'harmonie, et pour une trompette-trombone, Au sieur André-Antony Schmittschneider, à Paris." In *Description des machines* XIII (Paris, 1827), 286–90; "Brevet d'invention 1849, 24 mars 1821, Pour des instruments de musique à vent et à clef, Au sieur Asté dit Halary (Jean-Hilaire), professeur de musique et facteur d'instruments métalliques à vent, à Paris." In *Description des machines* XXI (Paris, 1831), 197–202.

69. "Brevet d'invention 1752, 23 juin 1825, Pour un mécanisme qui s'adapte à la flûte traversière, Au sieur Delavena, marchand-luthier, à Lille, département du Nord." In *Description des machines* XX, (Paris, 1830), 120–1.

70. "Brevet d'invention 892, 18 mai 1818, Pour un nouveau cor et une nouvelle trompette", *Description des machines* X, (Paris, 1825), 213–14; "Brevet d'invention 150, 6 décembre 1803, Pour des moyens de construire des orgues, aux sieurs Philippe et Frédéric Girard frères, à Paris." In *Description des machines* II (Paris, 1818), 265–71.

71. "Brevet d'invention 2981, 24 novembre 1830, Pour un piano sans cordes, avec l'addition d'un moyen proper à en prolonger le son, Au sieur Pape (Henri), facteur de pianos, à Paris." In *Description des machines* XXXI (Paris, 1836), 192–3; "Brevet d'invention 2649, 11 septembre 1829, Pour un instrument de musique appelé piano éolien, Au sieur Kayser, à Strasbourg,

département du Bas-Rhin." In *Description des machines* XXVIII (Paris, 1836), 273–5.

72. On the theory, "Sur la différence de masses dans les deux sexes," see Antide Magnin, *Notions mathématiques de chimie et de médecine ou Théorie de feu* (Paris: Fuchs, 1900), 195–9.

73. "Brevet d'invention 701, 22 janvier 1816, Pour des orgues expressives, Au sieur Grenié." In *Description des machines* IX (Paris, 1824), 53–66. Voir "L'orgue expressif, Rapport à monsieur l'Intendant général de l'Argenterie, des Menus Plaisirs et Affaires de la Chambre du Roi par l'Inspecteur Général," 28 Nov. 1816, AJ/37 Archives de l'Ecole royale de musique et de déclamation, des Conservatoire impériaux, nationaux ou royaux de musique, ou de musique et de déclamation, à Paris (1784–1925)/3. AN.

74. "Brevet d'invention 696, 14 septembre 1815, Pour une espèce de chronomètre en usage dans la musique, Au sieur Maelzel."In *Description des machines* IX (Paris, 1824), 15–16; "Brevet d'invention 1813, 24 novembre 1815, Pour un instrument appelé métronome perfectionné, Au sieur Bien-Aimé Fournier, horloger, à Amiens, département de la Somme." In *Description des machines* XXI (Paris, 1831), 91–5; "Brevet d'invention 2852, 9 mai 1829, Pour des perfectionnemens apportés dans la construction du métronome à l'usage de la musique, Au sieur Maelzel, mécanicien, aux Etats Unis, faisant élection de domicile à Paris." In *Description des machines* XXVIII (Paris, 1836), 80–2.

75. "Brevet d'invention 2417, 17 janvier 1829, Pour un nouveau diapason appelé typotone, au sieur Pinsonnat, contrôleur de garantie de la marque d'or et d'argent, à Amiens, département de la Somme." In *Description des machines* XXVI (Paris, 1835), 344–5; "Brevet d'invention 2829, 18 mai 1830, Pour un instrument appelé accordeur, Au sieur Salomon (Jean-François), professeur de musique à Besançon, département du Doubs, faisant élection de domicile à Paris." In *Description des machines* XXX (Paris, 1836), 66–7.

76. "Brevet d'invention 1496, 3 juillet 1823, Pour un mécanisme appelé tourne-feuille, Au sieur Puyroche, de Genève." In *Description des machines* XVII (Paris, 1829), 26–7.

77. "Brevet d'invention 2918, 10 novembre 1827, Pour une boite mélotachygraphique servant à fondre les planches propres à la gravure de la musique, Au sieur Petitpierre, ingénieur-mécanicien, à Paris." In *Description des machines* XXX (Paris, 1836), 340–1; "Brevet d'invention 478, 28 novembre 1801, Pour des procédés propres à imprimer la musique avec la presse typographique, Aux sieurs Duplat et George, à Paris." In *Description des machines* VI (Paris, 1824), 309.

78. "Brevet d'invention 1051, 24 janvier 1820, Pour des cartelles à l'usage de la musique, Au sieur Ferdinand Adrien, à Paris." In *Description des machines* XI (Paris, 1825), 306–7.

79. "Brevet d'invention 2878, 10 novembre 1830, Pour un appareil en forme de pupitre de musique, appelé transpositeur musical, Au sieur Leullier, à Paris." In *Description des machines* XXX (Paris, 1836), 192–4.

80. "Brevet d'importation 964, 12 mars 1819, Pour un mécanisme appelé chiroplaste, Au sieur Galliani et Serré, à Paris." In *Description des machines* XI (Paris, 1825), 23–6.

81. See James Kennaway's work, in particular, his article, "Musical Hypnosis: Sound and Selfhood from Mesmerism to Brainwashing," *Social History of Medicine* (2011), doi:10.1093/shm/hkr143.

82. "Aucune classe de société n'est exposée aussi fréquemment aux troubles nerveux en général et aux troubles cérébraux tout particulièrement. La céphalalgie, la migraine, les étourdissements, le vertige, l'insomnie, les perturbations sensorielles, l'irritabilité générale, l'hypochondrie, la mélancolie, sont les phénomènes que l'on observe chez eux." Maurice Krishaber, "Musicien (Hygiène des)." In *Dictionnaire encyclopédique des sciences médicales* ed. Amédée Dechambre, Deuxième Série, vol. 11 (Paris: Masson, 1875), 129–32.

83. Maurice Krishaber, *De la névropathie cérébro-cardiaque* (Paris: Masson, 1873).

84. J.-A. Barral (ed). *Œuvres complètes de François Arago, Mémoires scientifiques 2*, vol. 11 (Paris: Gide, 1859), 1–12. This experiment and many others at this time are also recounted in a number of the popular science books such as, Jules Gavarret, *Le Son: Notions d'acoustique physique musicale* (Paris: Hachette, 1975).

85. "Il ne faut pas croire que la cadence et l'harmonie des exercices militaires soient un pur objet de luxe et de parade; elles produisent les effets les plus précieux. Elles habituent le soldat à régler tous ses mouvements sur la voix de ses chefs, et sur le son des instruments guerriers; elles rendent un de ses organes plus docile aux impressions sonores, et par conséquent plus susceptible d'exaltation et d'entraînement, lorsqu'on voudra produire avec ensemble un grand résultat d'action." Charles Dupin, *Géométrie et mécanique des arts et métiers et des beaux-arts: cours normal à l'usage des artistes et des ouvriers, des sous-chefs, et des chefs d'ateliers et des manufactures*, vol. 3 (Bruxelles: Mat Fils et H. Rémy, 1825), 43.

86. "Brevet d'invention 1792, 27 décembre 1828, Pour une langue musicale pouvant être employée comme telégraphe musicale nocturne et comme langage secret, Au sieur Sudre, compositeur de musique, à Paris." In *Description des machines* XX (Paris, 1830), 360–64.

87. Ibid., "Nous avons donc pensé qu'offrir aux hommes un nouveau moyen de se communiquer leurs idées, de se transmettre à des distances éloignées et dans l'obscurité la plus profonde, était un véritable service rendu à la société."

88. Théodore Lachez, *Acoustique et optique des salles de réunion publiques, théâtres et amphithéâtres, spectacles, concerts, etc. suivies d'un projet de salle d'assemblée constituante pour neuf cents membres* (Paris: l'auteur, 1848).

89. Such apparatus are powerful metaphors for the unique model of French industrialization that has been described in recent economic history studies. Economic historians such as Jeff Horn demonstrate that French industrialization was built on the complex intervention of the state, forced to accommodate workers' interests and foster niche markets of highly specialized forms of technical expertise. State policy was not able to repress workers, as happened in England for the purposes of open competition and mass manufacturing. Rather, France constructed its own form of slow-building industrialization through the humane hearing model. State authority depended on workers' voices and technical input, and

these became hallmarks of industrial design success that slowly built manufacturing and industry.

90. Table XXVI, Adolphe de Pontécoulant, *Organographie: Essai sur la Facture Instrumentale – Art, Industrie et Commerce*, vol. 2 (Paris, 1861), 648.

91. See the Chapter 3 in Malou Haine, *Les facteurs d'instrument à Paris au XIXe siècle: des artisans face à l'industrialisation* (Bruxelles: Editions de l'Université de Bruxelles, 1985).

92. More specifically organ "Wind Instruments," *instruments à vent, à réservoir d'air et a tuyaux* ("wind instruments, with water reservoir and with pipes") and organ "Reed Instruments," *instruments à vent a réservoir d'air avec lames vibrantes ou avec cordes* ("Wind Instruments With Air Reservoir With Vibrating Blades or With Strings").

93. Haine, *Les facteurs d'instrument à Paris*, 145–6.

94. Ibid., 102.

95. See Viardot's purchase of a Cavaillé-Coll orgue expressif in Mark Everist, "Enshrining Mozart: *Don Giovanni* and the Viardot Circle," *Nineteenth-Century Music* 25, nos.2/3 (Fall/Spring 2001/2002), 165–89.

96. See for example, the advertisement for the *Harmonicorde* in *La Revue et Gazette Musicale de Paris* 2 March 1856.

97. The full French text from the review is the following "Nous avons gardé pour le dernier le compte-rendu du concert donné le 28 février par Mlle Judith Lion, d'abord parce qu'il a été un des plus brillants de la saison, et ensuite parce que la manière dont elle a fait valoir l'*harmonicorde* de Debain a fait événement. C'est la première fois qu'on entendait cet instrument en public. La combinaison de l'unicorde et de l'anche ou lame vibrante prête à l'harmonicorde des effets d'une grande nouveauté et d'une puissance extraordinaire qui tiennent de la harpe et du piano. Ce qui le distingue surtout, aussi bien, au reste, que les harmoniums de Debain, c'est une variété de timbres qui ne se rencontre dans aucun des instruments du même genre. L'imitation des divers instruments de l'orchestre y est poussée à un point qu'on ne dépassera plus et qui a déjà fait nommer l'harmonicorde l'*orchestre des salons.*" Concerts matinées et soirées musicales – Mlle Juith Lion. *La Revue et Gazette Musicale de Paris* (2 mars 1856), 69.

98. "la réunion de plusieurs instruments divers réveillait la sensibilité et la dirigeait: que les affectations de l'âme se trouvaient douées d'une nouvelle vie par les effets variés de plusieurs instruments savamment combinés: mais une réunion de plusieurs instruments divers étant presque toujours fort difficile à récontrer, on chercher un instrument qui à lui seul pût remplacer tous." Adolphe de Pontécoulant, *Brevet d'invention: Harmonium Debain* (Paris: Lacrampe, n.d.).

99. "Le piano manque la tenue des sons; le pianiste, privé de ce pouvoir magi que mystérieux qui pénètre si bien replus les plus profons du coeur, désespéré de son impuissance et la combattant en vain, cherche par d'autres moyens à impressionner le public. Ibid.

100. Cyril Ehrlich, *The Piano – A History* (London 1976), 121.

101. "Les claviers électriques, toujours entourés de nombreux auditeurs, sont l'objet d'une grande curiosité, chacun veut se rendre compte de l'effet produit par ces merveilleuses transmissions et on applaudit à cette nouvelle

victoire de la science." (from *Le Monde musical* [15 septembre 1889], 3, quoted in Michel Jurine, *Joseph Merklin, facteur d'orgues européens: Essai sur l'orgue français au XIXe siècle*, vol. 2 (Paris, 1991), 105.
102. Jurine, *Joseph Merklin*, 415.

## Conclusion

1. "J'avoue qu'il est difficile de comprendre comment cela se fait: ce sont des mouvements mécaniques qui sont imperceptibles, et dont il est très-difficile d'expliquer la nature et les causes." Duverney, *Traité de l'ouïe*, 81–2.
2. Nancy, *Listening*, 16.
3. I take the term "Imagined" here from Benedict Anderson, *Imagined Communities*, London: Verso, 2006.
4. Paul Fuchs et al., *Oxford Handbook of Auditory Science*, 3 vols (NY: Oxford, 2010).
5. Fuchs, ed. *Oxford Handbook of Auditory Science*, vol. 1, 1.
6. Rees and Palmer, *Oxford Handbook of Auditory Science*, vol. 2, 1.
7. Maria Chait, Alain de Chevigné, David Poeppel and Jonathan Z. Simon, "Neural Dynamics of Attending and Ignoring in Human Auditory Cortex," *Neurophyschologia* 48 (2010), 3262–71. doi:10.1016/j.neuropsychologia.2010.07.007.
8. Alain de Cheveigné and Jonathan Z. Simon, "Sensor Noise Suppression," *Journal of Neuroscience Methods* 168 (2008), 201. doi:10.1016/j.jneumeth.2007.09.012201.
9. "The Elephant Listening Project," The Elephant Listening Project, accessed 19 April 2013: Available at: http://www.birds.cornell.edu/brp/elephant/index.html.http://www.birds.cornell.edu/brp/elephant/index.html.
10. "What Does a Glacier Sound Like?" accessed 19 April 2013: Available at: http://www.guardian.co.uk/environment/blog/2011/sep/19/what-does-glacier-sound-like.
11. Lowery, David, Virginia Gray and Frank Baumgartner, "Policy Attention in State and Nation: Is Anybody Listening to the Laboratories of Democracy," *Publius: The Journal of Federalism* 41, no. 2, 286. doi: 10.1093/publius/pjq039.
12. James Adams, Lawrence Ezrow and Zeynep Somer-Topcu, "Is Anybody Listening? Evidence That Voters Do Not Respond to European Parties' Policy Statements During Elections," *American Journal of Political Science* 55, no. 2 (April 2011), 371.
13. Japhet Ezra July Mchakulu, "Youth Participation in Radio Listening Clubs in Malawi," *Journal of Southern African Studies* 33, no. 2 (June 2007), 251–65.
14. Julian Roberts, "Listening to the Crime Victim: Evaluating Victim Input at Sentencing and Parole," *Crime and Justice* 38, no. 1 (2009), 356.
15. Ibid., 381.
16. Katherine Schultz, Cheryl E. Jones-Walker and Anita P. Chikkatur, "Listening to Students, Negotiating Beliefs: Preparing Teachers for Urban Classrooms," *Curriculum Inquiry* 38, no. 2 (2008), 161. doi: 10.1111/j.1467/873X.2007.00404.x.
17. Ibid., 180.

18. Thijs Fassaert, Sandra van Dulmen, François Schellevis and Jozien Bensing, "Active Listening in Medical Consultations: Development of the Active Listening Observation Scale (ALOS-global)," *Patient Education and Counseling* 68 (2007), 258. doi: 10.1016/j.pec.2007.06.011.

19. Ibid., 259.

20. See, for example, "Tinnitus and Hypercusis Center: Professor Pawel J. Jastreboff," accessed 19 April 2013: Available at: http://www.tinnitus-pjj.com.

21. Hsu, Suh-Meei et al., "Associations of Exposure to Noise with Physiological and Psychological Outcomes among Post-cardiac Surgery," *Clinics* 65, no. 10 (October 2010), 985–9. PMCID: PMC2972598. Noise has also been identified as a significant cause of death in European urban environments by the World Health Organization (WHO). See Bolte, Gabriel, "Environment-Related Inequalities." In *Environmental Health Inequalities in Europe* by WHO European Centre for Environment and Health, (Copenhagen: World Health Organization, Europe, 2012), 86–113.

22. J. F. Byers and K. A. Smyth, "Effect of Music Intervention on Noise Annoyance, Heart Rate and Blood Pressure in Cardiac Surgery Patients," *American Journal of Critical Care* 6, no. 3 (May 1997), 183–91. PMID: 9131197.

23. See "Hearing Loss and Dementia Linked in Study: Release Date 14 February 2011," accessed 19 April 2013: Available at: http://www.hopkinsmedicine.org/news/media/releases/hearing_loss_and_dementia_linked_in_study.

24. On the use of medical technologies see for example Robert Evans and Alexandra Plows, "Listening without Prejudice: Re-Discovering the Value of the Disinterested Citizen," *Social Studies of Science* 37, no. 6 (December 2007), 827–53. doi: 10.1177/0306312707076602.

# Bibliography

## Manuscript Sources

*Archives Nationales de France* (AN)
Châtelet de Paris, Chambre de l'Auditeur Y 8248–8406, Minutes de sentences. 1691–1791. Y 8407, Sentences et procès-verbaux divers, 1690–1738.
Grande Chancellerie et Conseil V¹520, G-J, 1785, Lettres de provision d'offices.
Archives Imprimées du pouvoir législatif de la Révolution à la IVe République: Impression des Assemblées AD XVIII C 69, 1789–1795, Procès-Verbal de l'Assemblée Nationale.
Textes administratifs classés méthodiquement (XIIIe-XIXe s.): deuxième partie de la collection dite Rondonneau. AD II 7, Mémoire pour justifier le droit du Sceau du Châtelet, attributive zde Jurisdiction par tout le Royaume. Paris, 1437.
Plans et dessins d'architecture: Paris et le département de la Seine. XVIe–XIXe siècles. N/III/Seine/476/1-3, Châtelet de Paris: projets d'agrandissement, 3p. pl. et élév. Bruand, 1676.
Les archives des assemblées nationales C//167, no. 149. Assemblée nationale legislativ. Petition à l'Assemblée Nationale des Aveugles suivants l'Institution de M. Haüy. Paris, 1761.
Conseil exécutif et Convention AF*II 39, folio 11. Comité des secours publics, Procès-Verbaux 1791–an IV.
Hospices et Secours: Instituts nationaux – Aliénés F 15 2581, Etablissement national des Quinze-Vingts: Règement de Police de l'Intérieur de l'établissement. Paris, 1800.
Archives de l'Ecole royale de musique et de déclamation, des Conservatoire impériaux, nationaux ou royaux de musique, ou de musique et de déclamation, à Paris. 1784–1925. AJ/37/3, L'orgue expressif, Rapport à monsieur l'Intendant général de l'Argenterie, des Menus Plaisirs et Affaires de la Chambre du Roi par l'Inspecteur Général. Paris, 1816.

*Bibliothèque Nationale de France: Archives et Manuscrits* (BNF AM).
Français 21570–21574, Châtelet, Collection formée par Nicolas DELAMARE sur l'administration et la police de Paris et de la France.
*Hommage rendu de M. Poyet.* Signé B. Paris, 1824.

*Bibliothèque Nationale de France: Estampes et Photographie* (BNF EP).
*Plan du Châtelet de Paris.* Robert de Cotte, 1685.
*Grand concert extraordinaire exécuté par un détachement des quinze-Vingts au Caffé des Aveugles, Foire Saint Ovide au Mois de Septembre 1771.* Paris: Chez Mondhare rue St Jacques, n.d.

*Musée Valentin Haüy, Paris, France.*
*Troisième note du citoyen Haüy, auteur de la manière d'instruire les aveugles ou court exposé de la naissance des progrès et de l'état actuel de l'Institut national des*

*aveugles-travailleurs au 19 brumaire an IX de la République française* [10 November 1800] *entremêlée de quelques observations relatives à cet établissement.* Paris, 1800.
Haüy, Valentin. *Lettre à Le Sueur.* Paris, le 6 juillet, 1804.
1789, feuillets signes musicaux.
Epreuves des caractères des aveugles gravés et fondus pour être imprimés en relief. Paris, n.d.

## Other Sources

Adams, James. Lawrence Ezrow and Zeynep Somer-Topcu. "Is Anybody Listening? Evidence That Voters Do Not Respond to European Parties' Policy Statements During Elections," *American Journal of Political Science* 55, no. 2 (April 2011): 370–82.

Andrews, Richard Mowery. *Law Magistrancy and Crime in Old Régime Paris, 1735– 1789.* Cambridge: Cambridge University Press, 1994.

Attali, Jacques. *Noise: The Political Economy of Music.* Translated by Brian Massumi. Minneapolis: University of Minnesota Press, 2009.

Barral, Jean-Augustin, ed. *Œuvres complètes de François Arago, Mémoire Scientifiques 2. Vol. 11.* Paris: Gide, 1859.

Békésy, Georg v. and Walter A. Rosenblith. "The Early History of Hearing – Observation and Theories," *The Journal of the Acoustical Society of America* 20, no. 6 (November 1948): 727–48.

Bernauer, James and David Rasmussen, eds. *The Final Foucault.* Cambridge: MIT Press, 1987.

Bichat, Xavier. *Recherches physiologiques sur la vie et sur la mort.* 3rd ed. Paris: Brosson, 1805.

Biget, Michelle. *Musique et Révolution française.* Paris: Les Belles Lettres, 1989.

Bimbenet-Privat,   Introduction:   *Série   Y,*   Available   at:   http://www .archivesnationales.culture.gouv.fr/chan/chan/pdf/sa/Y-0-Intro.pdf.

Boisseau, Eugène. "Hôpitaux, Hospices." In *Dictionnaire encyclopédie des sciences médicales. Quatrième série. Vol. 14, 275–425.* Paris: G. Masson et P. Asselin, 1888.

Bolte, Gabriel. "Environment-Related Inequalities." In *Environmental Health Inequalities in Europe by WHO European Centre for Environment and Health,* 86–113. Copenhagen: World Health Organization, Europe, 2012.

Bordeu, Théophile. de. *Recherches sur le pouls par rapport aux crisis. 3 Vols.* Paris: Didot le jeune, 1768–1772.

Born, Georgina. *Rationalizing Culture: IRCAM, Boulez and the Institutionalization of the Avant-Garde.* Los Angeles: University of California Press, 1995.

Bregman, Albert S. *Auditory Scene Analysis: The Perceptual Organization of Sound.* Cambridge, MA: MIT Press, 1990.

Brockliss, Laurence and Colin Jones. *The Medical World of Early Modern France.* Oxford: Clarendon Press, 1997.

Buisson, Mathieu-François-Régis. *De la division la plus naturelle des phénomènes physiologiques considérés chez l'homme.* Paris: 1802.

Bull, Michael and Les Back, eds. *The Auditory Culture Reader.* Oxford: Berg, 2003.

Byers, J. F. and K. A. Smyth. "Effect of Music Intervention on Noise Annoyance, Heart Rate and Blood Pressure in Cardiac Surgery Patients," *American Journal of Critical Care* 6, no. 3 (May 1997): 183–91. PMID: 9131197.

Caradonna, Jeremy. "The Monarchy of Virtue: The Prix de vertu and the Economy of Emulation in France, 1777–1791," *Eighteenth-Century Studies* 41, no. 4 (Summer 2008): 443–58.

Chait, Maria, Alain de Chevigné, David Poeppel and Jonathan Z. Simon. "Neural Dynamics of Attending and Ignoring in Human Auditory Cortex," *Neurophyschologia* 48 (2010): 3262–71. doi:10.1016/j.neuropsychologia.2010.07.007.

Chambre des Pairs. *Observation sur le projet de loi relatif à la suppression des Conseilleurs-Auditeurs.* Paris: Duverger, n.d.

Chassin, Charles-Louis. *Les Elections et les Cahiers de Paris en 1789: Documents Recueillis mis en ordre et annotés. Vol. 3.* Paris: Jouraust et Sigaux, Charles Noblet, Maison Quantin, 1885.

——. *Les Elections et les Cahiers de Paris en 1789: Documents Recueillis mis en ordre et annotés. Vol. 2.* Paris: Jouraust et Sigaux, Charles Noblet, Maison Quantin, 1885.

Chéreau, A. "Du Verney." In *Dictionnaire encyclopédie des sciences médicales. Première série. Vol. 30,* edited by Amédée Dechambre, 729–31. Paris: G. Masson et P. Asselin, 1884.

Cheveigné, Alain de. "Pitch Perception Models – A Historical Review," *Paper Presented at International Conference on Acoustics,* Kyoto, 2004.

Cheveigné, Alain de and Jonathan Z. Simon. "Sensor Noise Suppression," *Journal of Neuroscience Methods* 168 (2008): 195–202. doi:10.1016/j.jneumeth.2007.09.012.

Clarke, Edwin and L. C. Jacyna. *Nineteenth-Century Origins of Neuroscientific Concepts.* Berkeley: University of California Press, 1987.

Cohen, Albert and Leta Miller. *Music in the Paris Academy of Sciences, 1666–1793. Detroit Studies in Music Bibliography* no. 43. Detroit: Information Coordinators Inc., 1979.

"Concerts matinées et soirées musicales – M^lle Judith Lion," *La Revue et gazette musicale de Paris* (2 mars 1856): 69.

Condillac, Etienne Bonnot de. *Essai sur l'origine des connaissances humaines, ouvrage où l'on réduit à un seul principe tout ce qui concerne l'entendement humain. 2 Vols.* Amsterdam: Mortier, 1746. Translated by Mr Nugent as *An Essay on the Origin of Human Knowledge* being a Supplement to Mr Locke's Essay on *Human Understanding.* London: Nourse, 1756.

Corbin, Alain. *Village Bells: Sound and Meaning in the French Countryside.* Translated by Martin Thom. New York: Columbia University Press, 1998.

Crevier, Jean-Baptiste-Louis. *Rhétorique française. Vol. 1.* Paris: Saillant, 1767.

*Déclaration du Roi ... Augmentation de pouvoir et droits aux Auditeurs desdits Châtelet.* Paris, 1683.

*Déclaration du Roi qui ordonne les Auditeurs du Châtelet de Paris jouiront les Droits ... Paris,* 1572.

Deffaux, Marc. *Commentaire sur les Justices de Paix.* Paris: Cotillon, 1838.

"De la division la plus naturelle des phénomènes physiologiques considérés chez l'homme par M. Fr. R. Buisson." In *Bibliothèque Médicale ou Recueil Périodique: D'extraits des meilleurs ouvrages de médecine et de chirurgie* 1, no. 1, 26–44. Paris, 1803.

*Description des machines et procédés specifiés dans les brevets d'invention, de perfectionnement et d'importation. Vols. II–XXVIII.* Paris: Madame Huzard/Imprimerie Nationale, 1818–1837.

Desessartz, Jean-Louis. *Réflexion sur la musique considérée comme moyen curatif.* Paris, 1802.

——. *Résultats des observations faites dans plusieurs départements de la République, sur les maladies qui ont régné pendant les six premiers mois de l'An VIII.* Paris: Huzard, An VIII [1799–1800].

Desmaze, Charles. *Le Châtelet de Paris.* Paris: Didier, 1870.

"Dissertations sur l'organe de l'ouie ... par M. Geoffroy. A Paris ... 1778," *Histoire de la Société Royale de Médecine, année 1776, 95–7.* Paris: Philippe-Denys Pierres, 1779.

Dufey, Pierre-Joseph-Spiridion. *Des Auditeurs, ou Essai historique et critique sur les révolutions de l'ordre judiciaire en France, depuis l'introduction des auditeurs à la Cour des comptes ... jusqu'à nos jours.* Paris: Brissot, Thivars et Cie, 1828.

Duffin, Jacalyn. *To See with a Better Eye: A Life of R. T. H. Laennec.* Princeton: Princeton University Press, 1998.

Dupin, Charles. *Géométrie et mécanique des arts et métiers et des beaux-arts: cours normal à l'usage des artistes et des ouvriers, des sous-chefs, et des chefs d'ateliers et des manufactures. Vol. 3.* Bruxelles: Mat Fils et H. Rémy, 1825.

Dupont, Adrien. *Principes fondamentaux de la police et de la justice présentés au nom du comité de constitution, présenté par M. Dupont, Député de Paris.* Paris: Assemblée Nationale, 1791.

Duverney, Joseph-Guichard. *Traité de l'organe de l'ouïe.* Paris: Estienne Michallet, 1683.

Ehrlich, Cyril. *The Piano – A History.* London: Dent, 1976.

Emch-Dériaz, Antoinette. "De l'importance de tater le pouls," *Canadian Bulletin of Medical History* 18 (2001): 369–80.

Erlmann, Veit. *Reason and Resonance: A History of Modern Aurality.* New York: Zone, 2010.

Erlmann, Veit, ed. *Hearing Cultures: Essays on Sound, Listening and Modernity.* Oxford: Berg, 2004.

Evans, Robert and Alexandra Plows. "Listening without Prejudice: Re-Discovering the Value of the Disinterested Citizen," *Social Studies of Science* 37, no. 6 (December 2007): 827–53. doi: 10.1177/0306312707076602.

Everist, Mark. "Enshrining Mozart: Don Giovanni and the Viardot Circle," *Nineteenth-Century Music* 25, nos. 2/3 (Fall/Spring 2001–2002): 165–89.

*Extrait des Registres de l'Académie Royale des Sciences du 20 juin 1787, rapport des commissaires chargés, par l'Académie, des Projets relatifs de l'établissement des quatre Hôpitaux.* Paris: Imprimerie Royale, 1787.

*Extrait des Registres de l'Académie Royale des Sciences du 22 novembre 1786, rapport des commissaires chargés, par l'Académie, de l'examen du projet d'un nouvel Hôtel-Dieu.* Paris: Imprimerie Royale, 1786.

Farge, Arlette. *Vivre dans la rue à Paris au XVIIIe siècle.* Paris: Gallimard, 1992.

Fassaert, Thijs, Sandra van Dulmen, François Schellevis and Jozien Bensing. "Active Listening in Medical Consultations: Development of the Active Listening Observation Scale (ALOS-global)," *Patient Education and Counseling* 68 (2007): 258–64. doi: 10.1016/j.pec.2007.06.011

Fénelon, François. *Les aventures de Télémaque.* Paris: Vve. de C. Barbin, 1699.

———. *Réfutation du système du Père Malebranche sur la Nature et de la Grâce. In Œuvres philosophiques de Fénélon. Nouvelle collationnée sur les meilleurs textes.* Paris: Charpentier, 1843.

———. *Telemachus.* Edited by Patrick Riley and Translated by Patrick Riley. Cambridge: Cambridge University Press, 1994.

———. *Traité de l'existence et des attributs de Dieu. In Œuvres philosophiques de Fénélon. Nouvelle collationnée sur les meilleurs textes.* Paris: Charpentier, 1843.

Forgues, V. *De l'influence de la musique sur l'économie animale.* Montpellier: Martel, 1802.

Foucault, Michel. *Abnormal: Lectures at the Collège de France, 1974–1975.* Edited by Valerio Marchetti and Antonella Salmoni. Translated by Graham Burchell. New York: Picador, 2003.

———. *Death and the Labyrinth: The World of Raymond Roussel.* Translated by Charles Ruas. London: Continuum, 1986.

———. *Discipline and Punish: The Birth of the Prison.* London: Vintage, 1995.

———. *Dits et Ecrits.* Translated by Daniel Defert and François Ewald. Vol. 1. Paris: Gallimard, 1954.

———. "Introduction by Michel Foucault." Translated by Donald F. Bouchard and Sherry Simon. In *The Temptation of Saint Anthony* by Gustave Flaubert. Translated by Lafcadio Hearn, xxiii–xliv. New York: Modern Library, Random House, 2001.

———. *Of Other Spaces* (1967). Translated by Jay Miskowiec, Diacritics (1992), 16, 22–6. Available at: http://foucault.info/documents/heteroTopia/foucault .heteroTopia.en.html.

———. *The Birth of the Clinic.* Oxon: Routledge, 2003.

———. *The Order of Things: An Archaeology of the Human Sciences.* New York: Vintage Books, 1994.

Fuchs, Pauls, Alan Palmer, Adrian Rees and Chris Plack, eds. *The Oxford Handbook of Auditory Science. 3 Vols.* New York: Oxford, 2010.

Furet, François. *Interpreting the French Revolution.* Translated by Elborg Forster. Cambridge: Cambridge University Press, 1997.

———. *Introduction to the Old Regime and the Revolution,* by Alexis de Tocqueville. Vol. 1. Edited by François Furet and Françoise Mélonio. Translated by Alan Kahan, 1–79. Chicago: University of Chicago Press, 1998.

Gaillard, Gabriel-Henri. *Rhétorique française à l'usage des jeunes demoiselles avec des exemples tirés pour la plupart de nos meilleurs Orateurs & Poètes Modernes.* 5th ed. Paris: Veuve Savoie, 1776.

Gavarret, Jules. Le son: *Notions d'acoustique physique musicale.* Paris: Hachette, 1975.

Gell, Alfred. *Art and Agency: An Anthropological Theory.* Oxford: Clarendon Press, 1998.

Geoffroy, Etienne-Louis. *Dissertations sur l'organe de l'ouïe, 1. De l'homme, 2. Des reptiles, 3. Des poissons.* Paris: Cavalier, 1778.

Gerbaud, Henri and Isabelle Foucher. *Etat générale des fonds,* Available at: http://www.archivesnationales.culture.gouv.fr/chan/chan/fonds/EGF/SA/SAPDF/Egfn-y.pdf.

Gibson, Jonathan. " 'A Kind of Eloquence Even in Music': Embracing Different Rhetorics in Late Seventeenth-Century France," *The Journal of Musicology* 25, no. 4 (Fall 2008): 394–433.

Goldstein, Jan. *The Post-Revolutionary Self: Politics and Psyche in France, 1750–1850.* Cambridge, MA: Harvard University Press, 2005.

Goodman, Dena. "Public Sphere and Private Life: Towards a Synthesis of Current Historiographical Approaches to the Old Regime," *History and Theory* 31, no. 1 (February 1992): 1–20.

——. *The Republic of Letters: A Cultural History of the French Enlightenment.* Cornell: Cornell University Press, 1996.

Gouk, Penelope. *Music, Science and Natural Magic in Seventeenth-Century England.* New Haven: Yale University Press, 1999.

——. "Music and the Nervous System in Eighteenth-Century British Medical Thought," *Unpublished chapter,* 7 February 2012.

Habert, L. *Pratique du sacrament de penitence ou méthode pour l'administration utilement.* Paris, 1748.

Haine, Malou. *Les facteurs d'instrument à Paris au XIX^e siècle: des artisans face à l'industrialisation.* Bruxelles, Editions de l'Université de Bruxelles, 1985.

Hautesierck, François-Marie-Claude Richard de. *Nouveau plan de constitution pour la médecine en France presenté à l'Assemblée nationale par la Société Royale de Médecine.* Paris, 1790.

Haüy, René-Just. *Traité élémentaire de physique médicale.* 4th ed. Paris: Delahays, 1855.

Haüy, Valentin. *Adresse du citoyen Haüy, auteur des moyens d'éducation des enfans aveugles et leur premier instituteur aux 48 sections de Paris,* présentée à la suite d'une adresse de la section de l'Arsenal, en date de l'an I de la République française le 13 décembre 1792, dont il étoit porteur.

——. *Essai sur l'éducation des aveugles.* Paris: Imprimé par les Enfans-Aveugles, 1786.

——. *Nouveau syllabaire ... manuel de l'élève [et manuel de l'instituteur], par le citoyen Haüy.* Paris: Institut national des aveugles travailleurs, an VIII [1799–1800].

"Hearing Loss and Dementia Linked in Study: Release Date 14 February 2011," *John Hopkins Medicine: News and Publications.* Accessed 19 April 2013. Available at: http://www.hopkinsmedicine.org/news/media/releases/hearing_loss _and_dementia_linked_in_study.

Henri, Pierre. *La vie et l'œuvre de Valentin Haüy.* Paris: Presses Universitaires de France, 1984.

Hoffbauer. *Paris: A Travers les âges: Aspects successifs des monuments et quartiers historiques de Paris: Depuis le XVIIIe siècle à nos jours.* 2nd ed. Paris: Firmin-Didot, n.d.

Horn, Jeff. *The Path Not Taken: French Industrialization in the Age of Revolution, 1750–1830.* Cambridge, MA: MIT Press, 2006.

Horne, Jacques de, ed. *Journal de médecine militaire publié par ordre du Roi. Paris: Imprimerie Royale,* 1782–1789. Translated by Joseph Browne as *A Journal of the Practice of Medicine, Surgery and Pharmacy in the Military Hospitals of France 1. Vol. 1.* New York: McClean & Co., n.d.

Hsu, Suh-Meei et al. "Associations of Exposure to Noise with Physiological and Psychological Outcomes among Post-Cardiac Surgery," *Clinics* 65, no. 10 (October 2010): 985–9. PMCID: PMC2972598.

Hunt, Lynn. *Inventing Human Rights: A History.* New York: Norton, 1997.

——. *Politics, Culture and Class in the French Revolution*. Berkeley: University of Los Angeles, 1984.

Husson, Armand. *Etude sue les hôpitaux considérées sous le rapport de leur construction de la distribution de leurs bâtiments, de l'ameublement, de l'hygiène, et du service des salles de malades*. Paris: Dupont, Administration de l'Assistance Publique, 1862.

Johnson, James. *Listening in Paris: A Cultural History*. Los Angeles: University of California Press, 1996.

Jurine, Michel and Joseph Merklin. *facteur d'orgues européens: Essai sur l'orgue français au XIXe siècle. Vol. 2*. Paris, Aux Amateurs Livres, 1991.

Katherine, Schultz, Cheryl E Jones-Walker and Anita P. Chikkatur. "Listening to Students, Negotiating Beliefs: Preparing Teachers for Urban Classrooms," *Curriculum Inquiry* 38, no. 2 (2008): 155–87. doi: 10.1111/j.1467/873X.2007.00404.x.

Kennaway, James. "Musical Hypnosis: Sound and Selfhood from Mesmerism to Brainwashing," *Social History of Medicine Details* 25, no. 2 (2011): 271–89. doi:10.1093/shm/hkr143.

Klein, Lawrence. "Enlightenment as Conversation." In *What's Left of Enlightenment?: A Postmodern Question*, edited by Keith Michael Baker and Peter Hanns Reill, 148–66. Stanford: Stanford University Press, 2001.

Klonk, Charlotte. "Science, Art and Representation of the Natural World." In *Cambridge History of Science. Vol. 4, Eighteenth-Century Science*, edited by Roy Porter, 584–617. Cambridge: Cambridge University Press, 2003.

Krishaber, Maurice. "Musicien (Hygiène des)." In *Dictionnaire encyclopédie des sciences médicales. Deuxième série. Vol. 11*, edited by Amédée Dechambre, 129–32. Paris: G. Masson et P. Asselin, 1875.

——. *De la névropathie cérébro-cardiaque*. Paris: Masson, 1873.

Kuriyama, Shigehisa. *The Expressiveness of the Body and the Divergence of Greek and Chinese Medicine*. Illustrated edition. New York: Zone Books, 1999.

Lachez, Théodore. *Acoustique et optique des salles de réunion publiques, théâtres et amphithéâtres, spectacles, concerts, etc. suivies d'un projet de salle d'assemblée constituante pour neuf cents membres*. Paris: l'auteur, 1848.

Lamy, Bernard. *La rhétorique ou l'art de parler*. 4th ed. Paris: Pierre Debats and Imbert Debats, 1701 [1676]. Translated as *The Art of Speaking rendered into English*. London: Godbid, 1676.

——. *Nouvelles réflexions sur l'art poétique*. Paris, 1768.

Lamy, Guillaume. "Addition curieuse." In *Explication mécanique et physique des fonctions de l'âme sensitive*. Paris: Lambert Roulland, 1681.

——. "Préface," "De l'ouïe," "Et une Description exacte de l'Oreille." In *Explication mécanique et physique des fonctions de l'âme sensitive*. Paris: Laurent d'Houry, 1687.

——. *Discours anatomiques, Avec des Réflexions sur les Objections qu'on lui a faites contre sa manière de raisonner de la Nature de l'Homme, et de l'usage des parties qui le composent*. Rouen: Lucas, 1675.

——. *Explication mécanique et physique des fonctions de l'âme sensitive*. Paris: Lambert Roulland, 1678.

——. *Explication mécanique et physique des fonctions de l'âme sensitive*. Paris: Laurent d'Houry, 1683.

Laqueur, Thomas. "Bodies, Details and the Humanitarian Narrative." In *The New Cultural History*, edited by Lynn Hunt, 176–204. Berkeley: University of California Press, 1989.

Le Cat, Claude. *Traité du mouvement musculaire de la sensibilité et de l'irritabilité*. Berlin, 1765.

*Le dictionnaire de l'Académie française, dédié au Roy*. Vol. 1. Paris: Vve J. B. Coignard et J. B. Coignard, 1694.

Lettre de M. *Picard, avocat et juge auditeur, à MM. les membres de l'Assemblée nationale, formant le Comité de constitution*. Paris: Postillon, 1791.

*Lettres contenant quelque reflexions sur les abus de l'administration de la Justice à Paris & un premier moyen d'y rémedier, enrichie de bonnes notes*. London and Paris: Monmoro Librairie, 1789.

*Lettres patentes du roi Philippe le Bel portant defences aux auditeurs*. Saint-Denis, 1311.

Lindsay, R. Bruce. "The Story of Acoustics," *The Journal of the Acoustical Society of America* 39 (1966): 629–44.

Louat, André and Jean-Marc Servat. *Histoire de l'industrie française jusqu'en 1945: Une industrialisation sans révolution*. Rosny: Bréal, 1995.

Lowery, David, Virginia Gray and Frank Baumgartner. "Policy Attention in State and Nation: Is Anybody Listening to the Laboratories of Democracy," *Publius. The Journal of Federalism* 41, no. 2: 286–310. doi: 10.1093/publius/pjq039.

Luchaire, Achille. *Manuel des institutions français, période de capetiens directs*. Paris: Hachette, 1892.

Lyall, Sarah. "Canvas? Paint? No, Just Sound," *The New York Times* 10 December 2011. Accessed 19 April 2013. Available at: http://www.nytimes.com/2010/12/11/arts/design/11turner.html?_r=0.

Magnin, Antide. *Notions mathémathiques de chimie et de médecine ou Théorie de feu*. Paris: Fuchs, 1900.

Malebranche, Nicolas. "De la recherche et de la verité." Translated by Geneviève Rodin-Lewis. In *Œuvres complètes de Malebranche*. Paris: J. Vrin, 1958–1984.

———. *The Search after Truth*. Translated and Edited by Thomas M. Lennon and Paul J. Olscamp. Cambridge: Cambridge University Press, 1997.

Marquet, François-Nicolas. *Nouvelle méthode facile et curieuse pour connaître le pouls par les notes de la musique*. 2nd ed. Paris: Didot le Jeune, 1769.

Matton, Sylvie. *Trois médecine philosophiques de XVIIe siècle*. Paris: Honoré Champion, 2004.

McGowen, Randall. "Power and Humanity, or Foucault among the Historians." In *Reassessing Foucault: Power, Medicine and the Body*, edited by Colin Jones and Roy Porter, 91–112. London: Routledge, 1994.

Mchakulu, Japhet Ezra July. "Youth Participation in Radio Listening Clubs in Malawi," *Journal of Southern African Studies* 33, no. 2 (June 2007): 251–65.

*Mémoire sur la nécessité de transférer et reconstruire l'Hôtel-Dieu de Paris, suivi d'un projet de translation de cet hôpital proposé par le sieur Poyet*. Paris, 1785.

"Mémoire sur l'ouïe des poissons et sur la transmission des sons dans l'eau, 24 avril 1743." In *Histoire de l'Académie royale des sciences avec les mémoires de mathématiques et de physiques tirés des registres de cette Académie* (1746).

Menuret, Jean-Joseph. *Nouveau traité du pouls*. Paris: Vincent, 1768.

Mercier, Louis-Sébastien. *Tableau de Paris*. Edited by Jean-Claude Bonnet. Paris: Mercure de France, 1994.

Nancy, Jean-Luc. *Listening*. Translated by Charlotte Mandell. New York: Fordham University Press, 2007.

Nollet, Abbé. *Leçons de physique experimentale. Vol. 3*. Paris: Guerin, 1745.

*Nouveau stile du Châtelet de Paris . . . tant en matière civile, criminelle, que de police.* Paris, 1771.

*Observations pour les aveugles de l'hôpital des Quinze-Vingts. Sur le projet de décret du comité de secours de la convention nationale, pour la suppression de cet hospital.* Paris, 179–.

Ong, Walter, J. *The Presence of the Word: Some Prolegomena for Cultural and Religious History*. New York: Global Publications, 2000.

*Ordonnance de M. le juge auditeurs*. Paris, 1772.

Osborne, Thomas. "On Anti-Medicine and Clinical Reason." In *Reassessing Foucault: Power, Medicine and the Body*, edited by Colin Jones and Roy Porter. London: Routledge, 1994, 28–47.

Otter, Chris. *The Victorian Eye: A Political History of Light and Vision in Britain, 1800–1900*. Chicago: Chicago University Press, 2008.

Ozouf, Mona. *La fête révolutionnaire, 1789–1799*. Paris: Gallimard, 1976.

Pascal, Blaise. *Œuvres complètes*. Paris: Seuil, 1963.

——. *Pensées*. Translated by A. J. Krailsheimer. London: Penguin, 1995.

Pascal, Jean-Joseph. *De l'ouïe considérée dans ses rapports avec l'intelligence et la voix*. Paris: Didot-le Jeune, 1821.

Perrault, Claude. *Essai de physique; ou, Recueil de plusieurs traitez touchant les choses naturelles. Vol. 1*. Paris: Jean Baptise Coignard, 1680.

——. *Essai de physique; ou, Recueil de plusieurs traitez touchant les choses naturelles. Vol. 2*. Paris: Jean Baptise Coignard, 1680.

Petit, Marc-Antoine. *Essai sur la médecine du cœur*. Lyon: Garbier, 1806.

Pontécoulant, Adolphe de. *Brevet d'invention: Harmonium Debain*. Paris: Lacrampe, n.d.

——. Organographie: *Essai sur la facture instrumentale – Art, Industrie et Commerce. Vol. 2*. Paris, 1861.

Poyet, Bernard. *Projet de cirque nationale et de fête annuelles*. Paris: Migneret, 1792.

——. *Projets de places édifices à ériger pour la gloire et l'utilité de la République*. Paris, 1799–1800.

Pressnitzer, Daniel. "Auditory Scene Analysis: The Sweet Music of Ambiguity," *Frontiers in Human Neuroscience* 5, no. 158 (December 2011): 1–11. doi:10.3389/fnhum.2011.00158.

Quinlan, Sean. "Physical and Moral Regeneration after the Terror: Medical Culture, Sensibility and Family Politics in France, 1794–1804," *Social History* 29, no. 2 (May 2004): 139–64.

Rameau, Jean-Philippe. *Code de musique pratique, ou méthodes pour apprendre la Musique même à des Aveugles [ . . . ]*. Paris: Imprimerie Royale, 1761.

Recaldé, Abbé de. *Traité sur les abus qui subsistent dans les hôpitaux du Royaume, et les moyens propres à les réformer*. Paris: Barrois, 1786.

Rees, Adrian and Alan Palmer, eds. *Overview to the Oxford Handbook of Auditory Science: The Auditory Brain. Vol. 2*, no. 1–8. New York: Oxford, 2010.

*Réimpression de l'ancien moniteur. Vol. 1*. Paris: Plon Frères, 1850.

*Réimpression de l'ancien moniteur. Vol. 13*. Paris: Plon Frères, 1847.

Richmond, Phyllis Allen. "The Hôtel-Dieu of Paris on the Eve of the Revolution," *Journal of the History of Medicine and Allied Sciences* 16 (1961): 335–53.

Riskin, Jennifer. *Science in the Age of Sensibility: The Sentimental Empiricists in the Age of Sensibility.* Chicago: University of Chicago Press, 2002.

Roberts, Julian. "Listening to the Crime Victim: Evaluating Victim Input at Sentencing and Parole," *Crime and Justice* 38, no. 1 (2009): 347–412.

Roger, Joseph-Louis. *Traité des effets de la musique sur le corps humain.* Edited and translated from the Latin text by Etienne Sainte Marie. Paris: Brunot, 1803.

Rosenfeld, Sophia. "On Being Heard: A Case for Paying Attention to the Historical Ear," *The American Historical Review* 116, no. 2 (April 2011): 316–34.

——. *A Revolution in Language: The Problem of Sign in Late Eighteenth-Century France.* Stanford: Stanford University Press, 2001.

Rousseau, Jean-Jacques. *The Social Contract.* Hertsfordshire: Wordsworth, 1998.

Royal, Almanach. Paris: D'Houry et Debure, 1791.

Sainte-Marie, Etienne. *Nouveau formulaire médical et pharmaceutique.* Paris: Rey et Gravier, 1820.

Smith, Mark, ed. *Hearing History: A Reader.* Georgia: University of Georgia Press, 2002.

Spary, Emma. Utopia's Garden: *French Natural History from the Old Regime to the Revolution.* Chicago: Chicago University Press, 2000.

Sterne, Jonathan *The Audible Past: Cultural Origins of Sound Reproduction.* Durham, NC: Duke University Press, 2003.

Sterne, Jonathan, ed. *The Sound Studies Reader.* London: Routledge, 2012.

Street, Alice and Simon Coleman. "Introduction: Real and Imagined Spaces," *Spaces and Culture* 15, no. 1 (2012): 4–17.

——. "Affective Infrastructure: Hospital Landscapes of Hope and Failure," *Space and Culture* 15, no. 1 (2012): 44–56.

Sykes, Ingrid. "Sounding the Citizen Patient: The Politics of Voice in Post-Revolutionary France," *Medical History* 55, no. 4 (2011): 479–502; PMCID: PMC3199644.

——. *Women, Science and Sound in Nineteenth-Century France.* Frankfurt am Main: Peter Lang, 2007.

Tenon, Jacques. *Mémoires sur les hôpitaux de Paris.* Paris: Imprimerie de Ph.-D Pierres et Royez, 1788.

Teule, Jules-Charles. *De l'oreille: essai d'anatomie et de physiologie, précédé d'un exposé des lois de l'acoustique.* Paris: Gabon, 1828.

*The Elephant Listening Project.* Accessed 19 April 2013. Available at: http://www.birds.cornell.edu/brp/elephant/index.html.

Tuetey, Alexandre. *Répertoire général des sources manuscrites de l'histoire de Paris pendant la Révolution française Vol. 6.* Paris: Imprimerie nouvelle, 1902.

——. *Répertoire général des sources manuscrites de l'histoire de Paris pendant la Révolution française Vol. 7.* Paris: Imprimerie nouvelle, 1905.

——. *Répertoire général des sources manuscrites de l'histoire de Paris pendant la Révolution française. Vol. 8.* Paris: Imprimerie nouvelle, 1908.

——. *Répertoire général des sources manuscrites de l'histoire de Paris pendant la Révolution française. Vol. 10.* Paris: Imprimerie nouvelle, 1912.

——. *Répertoire général des sources manuscrites de l'histoire de Paris pendant la Révolution française. Vol. 11.* Paris: Imprimerie nouvelle, 1914.

Tuzin, Donald. "Miraculous Voices: The Auditory Experience of Numinous Objects," *Current Anthropology* 25, no. 5 (December 1984): 579–96.

Vieussens, Raymond. *Traité nouveau de la structure de l'oreille divise en deux parties.* Toulouse: Jean Guillemette, 1714.

——. *Traité nouveau de la structure et des causes du mouvement naturel du cœur.* Toulouse: Jean Guillemette, 1715.

Voltaire. "Prix de la Justice et de la Humanité." In *Œuvres de Voltaire. Vol. 1: XIV, Mélanges,* edited by M. Beuchot, 253–336. Paris: Levèvre, 1834.

——. "Republican Ideas: By Member of a Public Body." In *Voltaire: Political Writings,* edited by David Williams, 195–211. Cambridge: Cambridge University Press, 1994.

Wason, Robert W. "Musica Practica: Music Theory as Pedagogy." In *Cambridge History of Western Music Theory,* edited by Thomas Christensen, 46–77. Cambridge: Cambridge University Press, 2002.

Weiner, Dora B. *The Citizen Patient in Revolutionary and Imperial Paris.* Baltimore and London: John Hopkins University Press, 1993.

Weiner, Dora B. and Michael J. Sauter. "The City of Paris and the Rise of Clinical Medicine," *Osiris 2nd Series 18, Science and the City* (2003): 23–42.

Weygand, Zina. *The Blind in French Society: From the Middle Ages to the Century of Louis Braille.* Translated by Emily-Jane Cohen with a preface by Alain Corbin. Stanford: Stanford University Press, 2009.

*What Does a Glacier Sound Like?* Accessed 19 April 2013. Available at: http://www.guardian.co.uk/environment/blog/2011/sep/19/what-does-glacier-sound-like.

Williams, Elizabeth A. *The Physical and the Moral: Anthropology, Physiology and Philosophical Medicine in France, 1750–1850.* Cambridge: Cambridge University Press, 1994.

# Index

Note: The letter 'n' following locators refers to notes.

Printed in the United States
by Baker & Taylor Publisher Services